AF882687

Network Security: Engineers Survival Guide

(A Field Manual)

© 2024 Dennis Adjepong. All rights reserved.

ISBN Paperback: 978-1-964467-09-2

ISBN Hardback: 978-1-964467-10-8

LCCN: 2024916499

All text, images, graphics, and other content in this publication are the exclusive property of Dennis Adjepong. No part of this publication may be reproduced, distributed, or transmitted in any form or by any means, including photocopying, recording, or other electronic or mechanical methods, without the prior written permission of the copyright owner, except in the case of brief quotations embodied in critical reviews and certain other non-commercial uses permitted by copyright law.

This publication is protected under international copyright laws and treaties. Unauthorized use, reproduction, or distribution of any part of this publication is strictly prohibited and may result in legal action.

All content, including text, images, and other materials, is owned by Dennis Adjepong. Any resemblance to real persons, living or dead, or real events is purely coincidental.

Dedication

I would like to dedicate this book to all past, present, and future students of Cybersecurity Information System University Global Online Campus www.cisuniv.org. This book will serve as a guide to assist all students pursuing a career change in Cybersecurity, Computer Science, and Computer Engineering related fields. It will serve as a must-use to make your IT job interview successful and get the intended job.

Table of Contents

CHAPTER 1:

ACRONYMS	LIST OF ABBREVIATIONS
CSO	*Chief Security Officer*
CTO	*Chief Technical Officer*
AAA	*Authentication, Authorization, and Accounting*
RADIUS	*Remote Authentication Dial-In User Service*
TACACS+	*Terminal Access Controller Access-Control System Plus*
CIA	*Confidential Integrity Availability*
APTs	*Advanced Persistent Threats*
SCCM	*Microsoft System Center Configuration Manager*
SMTP	*Simple Mail Transfer Protocol*
CIO	*Chief Information Officer*

Chapter 1

Introduction to Security.

The CIA of Computer Security (The CIA Triad – Computer Security)

This chapter discusses foundation-level security concepts for new IT security professionals. Information security is crucial in organizations to protect information systems from unauthorized access and prevent data modification or destruction without approval. Many IT directors have become full-fledged executives, such as CIO-(chief investigation officer), CSO-(Chief Security Officer), or CTO-(Chief Technical Officer). The book focuses on information security, identifying risks and vulnerabilities, and eliminating them. It teaches readers to think like a hacker but act like an administrator. This chapter is essential for those aspiring to become security administrators

Security 101

Commonly asked on Job-Interview !

No security system is truly secure, and while there are ways to construct security precautions, there is no perfect scenario. Understanding this, we can focus on building a solid foundation for mitigating security risks.

The CIA of Computer Security

Commonly asked on Job-Interview !

Computers can be victims of covert operations, and IT professionals must adhere to three core principles of information security: confidentiality, integrity, and availability. These principles are collectively known as the CIA triad. Confidentiality involves preventing the disclosure of information to unauthorized persons, such as Social Security numbers, driver's license information, bank accounts, and passwords. Integrity ensures data is not tampered with and requires authorization before modification. Availability means data is obtainable regardless of storage, access, or protection, and should be available regardless of malicious attacks.

The AAA-(Authentication, Authorization, and Accounting) of computer security consists of authentication, authorization, and accounting. Authentication establishes a person's identity with proof and is confirmed by a system, such as a digital identity or biometric data. Authorization allows users to access certain data or areas of a building, determined by permissions, access control lists, time-of-day restrictions, and other login and physical restrictions. Accounting tracks data, computer usage, and network resources, often involving logging, auditing, and monitoring.

The AAA concept should be applied to any security plan and is also used in authentication protocols like RADIUS-(Remote Authentication Dial-In User Service) and TACACS+ (Terminal Access Controller Access-Control System Plus). The details of AAA are set in stone within several RFC documents.

The Basics of Information Security

Commonly asked on Job-Interview !

Information security is the protection of data and information systems from unauthorized access, unlawful modification, disruption, disclosure, corruption, and destruction. To be an effective security administrator, organizations must be aware of several basic types of threats, including malicious software, unauthorized access, system failure, and social engineering. To implement information security, organizations should create a proactive security plan that includes physical, technical, and administrative controls. These controls protect the confidentiality, integrity, and availability (CIA) of data.

To prevent and help recover from previous threats, organizations should provide user awareness through employee training, education, and online resources. Authentication is a preventative measure that can be broken down into five categories: something the user knows, something they have, something they are, something they do, and somewhere they are. By implementing these controls, organizations can protect against and recover from potential security breaches, even on a low budget.

Anti-malware software protects computers from malware and detects and removes them, such as antivirus and anti-spyware programs. Data backups enable data recovery after attacks or system failures. Fault-tolerant methods like RAID 1, 5, 6, and 10 are good preventative measures against hardware failure but may not offer protection from data corruption or erasure. Encryption involves changing information using a cipher to make it unreadable to only authorized users. Data removal goes beyond file deletion or formatting, involving clearing, purging, and destruction. Combining a well-thought-out security plan with strong individual security methods can effectively stop threats and recover from them quickly and efficiently. The strongest security plans combine multiple methods in a layering strategy called defense in depth, protecting data throughout its entire life cycle.

Think like a Hacker

To think like a hacker, one must understand the hacker's mind, making computer and network security a difficult proposition. Some malicious individuals choose to become hackers for various reasons, such as user advantage, greed, agenda, or just to cause mayhem. There are different types of hackers, including white hats, black hats, gray hats, blue hats, and elite hackers. White hats are non-malicious individuals who attempt to hack into a computer system before it goes live to test it, often with a contractual agreement with the owner of the resource to be hacked. Black hats are malicious individuals who attempt to break into computers and networks without authorization, such as identity theft, piracy, and credit card fraud. Gray hats risk breaking the law by attempting to hack a system and not doing anything malicious, while blue hats are individuals asked to attempt to hack into a system by an organization but are not employed. Elite hackers are the ones who first discover vulnerabilities, and only 1 out of 10,000 hackers' wear the Elite hat.

To protect your organization's computer systems, it is essential to be on the lookout for potential threats and to review logs often. Security precautions, staying updated on the latest attacks, and never underestimating the power and resilience of a hacker are crucial.

Threat Actor Types and Attributes

Commonly asked on Job-Interview !

The security arena has various types of attackers, including script kiddies, hacktivists, cybercriminals, and insiders. Script kiddies are unsophisticated individuals with limited knowledge of technology, using code from others and freely accessible on the internet. They can knowingly or unintentionally inflict significant damage to insecure systems. Hacktivists, a combination of hack and activist, can be inside or outside a company and have varying resources and funding. Cybercriminals may work independently or be part of organized crime groups, often with high levels of sophistication. Advanced persistent threats (APTs) are individuals or groups that carry out cyberattacks for governments and nation states, often with the highest level of resources and motivation. Companies should be prepared for both basic and advanced attacks, treating them with respect. It is crucial to understand these actors and their motivations, but it is essential to be prepared to prevent basic and advanced attacks.

CHAPTER 2:

ACRONYMS	LIST OF ABBREVIATIONS
UEFI	*Unified Extensible Firmware Interface*
BIOS	*Basic Input/output System*
SSH	*Secure Shell*
AV	*Antivirus*
RAT	*Remote Access Trojan*
PHP	*Hypertext Preprocessor*
FTP	*File Transfer Protocol*
MITM	*Man-in-the-Middle*
DDoS	*Distributed Denial of Service*
URL	*Uniform Resource Locator*
MBR	*Master Boot Record*
GPT	*GUID Partition Table*
SMTP	*Simple Mail Transfer Protocol*
BCC	*Blind Carbon Copy*

Chapter 2

Anti-Virus Software

Seen mostlv!

Data is the most crucial aspect of a computer, and its security is crucial against threats like viruses, Trojans, and spyware. Applications like antivirus, anti-spyware, or combination anti-malware can help protect against malware. Regular updates and implementation of these security tools can prevent most malicious software attacks.

To combat computer security threats, classify them, define their delivery methods, and discuss prevention and troubleshooting strategies, starting with malicious software, which is the most common and deadly threat.

Malicious Software Types

Commonly asked!

Malicious software, including viruses, worms, Trojan horses, spyware, rootkits, and adware, is designed to infect a computer system and potentially damage it without user consent. To protect against malware, define and categorize it, implement preventative measures, and locate and quarantine it if it manifests itself.

The exam requires understanding various types of malware, including viruses, spyware, and ransomware. While most people are familiar with viruses and have antivirus software, many remain confused about other types and their prevention. Computer professionals focus on fixing non-virus-related issues and training users.

Viruses

A computer virus is code that runs without the user's knowledge, infecting the computer when accessed and executed. It has reproductive capability and can spread copies throughout the computer. Computers need to be up and running to spread the virus, making them more susceptible to infections.

The Love Bug, a 2000 virus, infected 15 million computers with an attachment containing a text file disguised as a Visual Basic script. It caused $5 billion in damage and deleted files. It's crucial to educate users on email security.

You may come across a variety of virus types:

- **Boot sector**: The virus loads into memory after first loading onto the hard drive's first sector during computer boot-up.
- **Macro**: Typically embedded in documents and sent to people via email to get them to open the document and launch the virus.
- **Program**: Affects executable files.
- **Encrypted**: Encrypts itself using a basic cipher. The virus is composed of a tiny decryption module and an encrypted copy of the virus code, which helps prevent discovery. Each file that is infected can employ a different set of encryption keys, but often there is only one decryption code.
- **Polymorphic**: Expands upon the idea of an encrypted virus, but each infection modifies the decrypting module. In an effort to evade antivirus detection, it can so alter each time it is run.

- **Metamorphic**: This type of infection resembles polymorphism but, in an additional effort to evade detection, rewrites itself entirely each time it infects a new file.
- **Stealth**: Makes use of a number of strategies to elude detection by antivirus software.
- **Armored**: Prevents malware by deceiving antivirus software into believing it is situated in a different location than its true location. In essence, it may defy attempts by analysts to investigate its code by using a layer of security that it has against the person attempting to analyze it.

Multipartite

A cross between a boot and a program virus that targets the system files or boot sector initially, then moves on to target the other files on the system. A virus hoax is a scam where a user receives a message claiming their computer is infected by a nonexistent virus, often in an email or pop-up, to trick them into paying for unnecessary tech support or remote computer access.

Worms

A worm self-replicates to spread to other computers, exploiting security holes in operating systems and applications. They replicate to systems running the same applications on the network or the Internet. Worms don't require a carrier or explicit instructions, unlike viruses that require a carrier. Nimda, a worm that spread automatically through the Internet in 2001, caused widespread damage through network shares, mass emailing, and operating system vulnerabilities. Worms, sometimes without a payload, can disrupt network traffic and computer operations due to their self-replicating abilities, even if they don't contain harmful code.

Trojan Horses

Quick Hits!

Trojan horses, which appear to perform desirable functions, actually perform malicious functions. They can be downloaded without detection and transferred to computers via removable media, such as USB flash drives. Users experiencing slow system performance or antivirus alerts may be infected by a Trojan. Remote access Trojans (RATs), such as BackOffice, NetBus, and SubSeven, are dangerous software that allow attackers' higher administration privileges than system owners. They can scan for unprotected hosts and make changes when connected. Although not malicious, they are easy to download and manipulate. An attacker can control a target computer, turning it into a bot.

RATs, coded in PHP, enable remote access to websites, like the web shell, allowing attackers to control servers without user consent, often by cracking FTP passwords. RATs, or advanced persistent threats, persistently target specific entities like governments or corporations, often using large-scale adversaries. These groups, often consisting of hacker factions and global corporations, use vast resources.

Ransomware

Ransomware, also known as crypto-malware, is a devious type of malware that restricts computer system access and demands a ransom. It encrypts files and locks the system, requiring payment to regain access. Cryptoviral extortion, also known as cryptoviral extortion, is a common form of ransomware. CryptoLocker, a ransomware Trojan, uses an RSA public key to encrypt files, making them nearly impossible to decrypt. Ransomware attacks have grown steadily since 2013, affecting hundreds of thousands of computers worldwide annually.

Note: Occasionally, a user may unintentionally stumble into a phony website (also known as a pop-up site) that claims all of the user's files have been encrypted and that payment is needed to unlock them. The statement is often accompanied by an imposing emblem that appears to be from the government. However, a lot of these websites don't genuinely encrypt user files. In this instance, there is no actual harm done to the computer or files—we are just talking about plain old extortion. Pop-up blockers, phishing filters, and users exercising common sense while clicking on links found through searches can all be used to block these kinds of websites.

Spyware

Spyware is malicious software, often downloaded or installed without user consent, which collects user information. It can be simple code logging website access or a key-logger recording keystrokes. Spyware is often associated with advertising and mal-advertising, using internet-based advertising to distribute

NOTE

Rootkits

A rootkit is a software designed to gain administrator-level control over a computer system without detection. It performs malicious operations on a target computer without the administrator's knowledge. Rootkits can target lower levels of the operating system, such as the boot loader and kernel. Examples include the Evil Maid Attack, which extracts encryption keys, and the Alureon rootkit, which affects the master boot record and low-level system drivers. Rootkits, activated before the operating system boots, can install hidden files, processes, and user accounts, intercepting data from network connections and keyboards.

malicious software. Spyware can alter computer settings without user interaction, redirecting browsers to inappropriate websites, and often pops up ads based on user spying. Grayware refers to improperly functioning applications without serious consequences. Spyware, such as Internet Optimizer, redirects error pages to advertising websites, can hijack computers and be used for cyber-espionage, as seen with Red October, which was installed when users were redirected to exploited PHP pages.

File-less Malware

The term "file-less" is a bit misleading because the attack may actually contain files. The affected systems, such as ATMs, frequently have weak endpoint security, and the organization that owns those systems may also have ineffective access control policies. Malware does not just have to reside on a computer's hard drive; it can also reside in RAM (and possibly other locations). File-less malware, also known as non-malware, functions without placing malicious executables within the file system, and instead works in a memory-based environment. Sound potent? It can be, but the attacker usually needs remote access to the system—via SSH, a RAT, or elsewhere.

Spam

Spam is the abuse of electronic messaging systems, such as email, texting, and social media, to send unsolicited bulk messages without benefiting the sender. Companies with questionable ethics condone this type of marketing, often set up as a pyramid scheme, to benefit the top of the marketing chain. Email spam is the most common form of spam, causing clogging, denial-of-service, and misleading users. Most network-based viruses are transferred through spam emails. Spimp, a derivative of spam, involves using instant messaging systems, chat rooms, and chat functions in games, causing legal issues for the organization that owns the email server.

Summaries of the Malware Threats

Key table!

Malware Threat	Definition	Example
Virus	Code that infects a computer when it is accessed and executed; it operates on the system without the user's awareness.	Love Bug virus Ex: love-letter-for-you.txt.vbs
Worm	Like viruses, except that it reproduce itself, unlike viruses.	Nimda Propagated through network shares and mass e-mailing
Trojan horse	Seems to carry out the intended tasks, while in reality, it is carrying out harmful operations in the background.	Remote access Trojan Ex: PlugX
Ransomware	Malware that locks users out of their computers and demands a ransom to be paid.	Often propagated via a Trojan Ex: CryptoLocker
Spyware	Malicious software can be installed in conjunction with other third-party software or can be inadvertently acquired from a website.	Internet Optimizer (a.k.a. DyFuCA)
Rootkit	Software made to covertly take over a computer system at the administrator level. Can target the kernel, boot loader, and UEFI/BIOS.	Boot loader rootkits Ex: Evil Maid Attack, Alureon

Spam	The misuse of broadcast media, instant chat, and email as electronic messaging platforms.	Identity theft e-mails (phishing) Lottery scam e-mails

Delivery of Malware

Malware is not sentient and requires transport and delivery to a computer. Attackers can gain physical access to an unprotected computer to perform their malicious work. Other methods include gaining access to a computer, making modifications, and delivering malware. The method used by a threat is known as a threat vector, and the means by which an attacker gains access to a computer for delivering malicious software is known as an attack vector.

Via Software, Messaging, and Media

Malware can be delivered through various methods, including email, FTP, peer-to-peer networks, bit torrents, websites, advertisements, and removable media. It can be easily embedded and distributed through corrupting code, bad downloads, or advertisements. Users should be cautious when using these technologies, as they can be manipulated to automatically run malware when inserted into a computer. Removable media can also contain hidden viruses, worms, and logic bombs to set the malware off at specific times.

Typo squatting or URL hijacking is a type of attack where attackers redirect users to malicious websites by squatting on similar but not exact domain names. Millions of typos occur daily in web browsers, making the system vulnerable to spyware and malware. While some browsers have built-in security features, training users to be cautious when typing domain names is the best way to protect against this.

Note: On typing, some attackers will use a key logger a hardware or software tool that records all keystrokes to capture all that a user types on the keyboard; security experts also frequently use key loggers; see Chapter 13, "Monitoring and Auditing" for more information.

Malware propagation involves using automation and web attack-based software kits, known as exploit kits, which target vulnerabilities in client computers. Examples include the Blackhole exploit kit, which is used by attackers to distribute malware to specific computers.

Malware can be distributed through a botnet, controlled by a master computer, where the individual compromised computers are called zombies. These zombies are unaware of the malware installed on them. Automated distribution of the malware can occur. If all zombie computers have a specific virus or attack loaded, a distributed denial-of-service (DDoS) attack can be enacted on any target. If a computer scans other systems, communicates with unknown servers, or has numerous outbound connections, it is likely part of a botnet.

Botnets can be used for fraud, such as the ZeroAccess botnet, which mines Bitcoins and perpetuates click fraud. It uses Trojan malware and is hidden from antivirus programs. The botnet, estimated to have 10 million computers, generates significant revenue. Despite being brought to justice, it is replaced by new entrepreneur each month.

Active Interception

Active interception involves capturing and potentially altering information between sender and receiver, potentially leading to theft, modification, or exploitation through methods like session theft and man-in-the-middle (MITM) attacks.

Privilege Escalation

Privilege escalation is the misuse of software bugs or design flaws to gain access to resources normally protected from the application, resulting in additional privileges for the user.

Backdoors

Backdoors are computer programs that bypass security mechanisms, originally used by developers for legitimate access. However, they have since been exploited by attackers to modify operating systems, websites, and network devices. They can also be created by worms for spamming or privilege escalation. Currently, backdoors are less common due to discouragement by software manufacturers and network device manufacturers. Network administrators should inform the manufacturer of any new backdoors.

Logic Bombs

Logic bombs are code inserted into software to initiate malicious functions when specific criteria are met. They blur the line between malware and a malware delivery system, activating viruses, worms, or Trojans at a specific time. Logic bombs can be contained within a virus or loaded separately. Systems administrators should notify their superiors and check their organization's policies if they suspect a logic bomb or a portion of its code. Actions may include network disaster recovery processes, notifying software vendors, and managing software usage.

Easter eggs, previously added as a joke, are now prohibited by software companies due to potential slipping past quality control. They can be found in Windows XP's Solitaire, and due to growing concerns about malware, companies have adopted Trustworthy Computing, a newer concept that sets standards for software design, code, and quality control. However, the Easter egg's day has passed in software.

Preventing and Troubleshooting Malware

Malware can cause system issues like slow response times, pop-ups, incorrect home pages, and unexpected system shutdowns. It often uses CPU and memory resources, causing slower performance. Technicians should look for erratic behavior from the computer. To prevent viruses and spyware, technicians should focus on preventing them and troubleshooting them if they occur.

Preventing and Troubleshooting Viruses

To protect a computer system from viruses, use strong endpoint protection platforms with antivirus software. AV software should be updated with a current license, which is usually renewed yearly. Regular full scans of the system within the software are recommended, either manually or automatically.

Antivirus systems detect viruses, worms, Trojans, logic bombs, rootkits, and botnet activity, but not logic bombs, rootkits, or botnet activity. AV software is essential but not a cure-all.

To ensure security, ensure the computer has the latest updates for both operating systems and applications. Enable Windows Update and check for updates periodically or set the system to automatically check for updates according to your company's policy. To ensure security, ensure the computer has the latest updates for both operating systems and applications. Enable Windows Update and check for updates periodically or set the system to automatically check for updates according to your company's policy.

A hardware-based firewall, like a SOHO router, provides two layers of protection against malware attacks. Set exceptions for programs requiring internet access, allowing specific applications to communicate through the firewall while keeping other ports closed. To prevent viruses, use "separation of OS and data" with two hard drives, with the operating system on the C: drive and data on the D: drive. This makes it harder for viruses to spread, easier to isolate, and allows easy reinstallation.

Note: Additionally, there are viruses that can harm different kinds of operating systems. Though they haven't historically been attacked as frequently as Windows computers, desktop operating systems like Linux, macOS, and Android are also vulnerable. Encryption is a crucial method to safeguard data from virus activity on a computer. Windows uses the Encrypting File System (EFS) to encrypt files individually, preventing clear-text access and requiring modification. While it may not prevent virus infections, it can protect individual files.

Educate users on virus infection, email screening, and scanning removable media. Consider hiring a technical trainer or creating interactive online tutorials for more effective learning. Providing an engaging learning environment can enhance user education and prevent system infections.

Using these techniques can significantly reduce virus infection, but it's crucial to identify the virus to effectively "cure" a computer.

The following are some common signs of viruses:

- The computer operates more slowly than normal.
- Computer frequently locks up or becomes unresponsive.
- Computer randomly restarts or crashes often.
- Applications, optical drives, and hard disks are inoperable or not functioning properly.
- There are odd noises.
- You get strange error warnings.
- There is distortion in the print or display.
- Old icons (and programs) vanish, or new ones appear.
- An email attachment that has been opened has two extensions; for instance, it may be.txt.vbs or.txt.exe.
- Installing or running antivirus software is not possible.
- Folders are generated automatically or files are corrupted.
- The ability to use System Restore has been blocked or eliminated.

The following steps are suggested for the general eradication of malware:

- Determine the signs of malware. 2. Systems that have been quarantined
- Turn off Windows' System Restore.
- Fix compromised systems: a. Upgrade anti-virus software.
- Make use of scan and removal methods (such as pre-installation environments and Safe Mode).
- Plan scans and execute upgrades.
- Make a restore point and activate System Restore (in Windows).
- Inform final consumers.

To prevent computer malware, back up critical data and ensure the latest updates are installed. Perform a thorough scan using AV software's scan utility in Safe Mode. Move the affected drive to a "clean machine" that is not connected to any network, and run the AV software on that machine to scan for malware.

AV software should detect and quarantine the virus on the system, but if it fails to find the issue or is infected, a free online scanner can be used.

In rare cases, deleting individual files and Registry entries may be necessary to remove a new virus without an antivirus definition, as instructions can be found on AV software manufacturers' websites.

AV software is the best option for boot sector viruses, as it can scan the boot sector using a bootable USB flash drive or disc, or with built-in boot shielding. Recovery environments and command-line can also be used.

Use Linux-based tools and Live Linux discs like Knoppix for computer booting and repair. Recovery environments may not fix the problem, so use AV software's utilities first

How to Avoid and Fix Trojan and Worm Issues

The same techniques that are used to stop viruses can also be used to stop worms and Trojan horses. Online Trojan scanners are available for free, however, most antivirus programs check for viruses, worms, and Trojan horses in addition to other types of malware. As previously discussed in the chapter, antivirus software can typically identify remote access Trojans with ease. This can be done by looking for the attacker's actual application or by looking for any.exe files that are used on the victim machine as part of the application.

Prevention involves maintaining and ensuring user interaction with computers, keeping AV software updated, and navigating legitimate websites and carefully screening emails.

Troubleshooting malware and ransomware involves quarantining or removing the malware using advanced techniques or AV software, as they are often delivered as Trojans.

Preventing and Troubleshooting Spyware

When it comes to employing a firewall and operating system updates, preventing spyware functions similarly to preventing viruses. Moreover, anti-virus providers and OS makers incorporate anti-spyware elements into their products because spyware is just as widespread as viruses.

Here are some additional steps you may take to safeguard your computer in the hopes of avoiding spyware:

- Update and use pre-installed anti-spyware apps like Windows Defender. Make sure the anti-spyware program is up to date.

- Modify the security settings of your web browser. For instance, limit or disable cookies, set up trusted zones, activate phishing filters, block undesired websites, enable automatic website scanning, disable scripting (such ActiveX and JavaScript), and make sure the browser clears its cache when it closes. Web-page spoofing, another term for fraudulent online demands for credit card numbers, usernames, and passwords, can be prevented with the aid of all these measures.

Session hijacking, the act of seizing control of a user session after getting or establishing an authentication ID, can also be prevented by adjusting security settings higher. We will discuss.

- Remove pointless programs and disable ostentatious services (such as FTP or Remote Desktop services, if not in use).
- Inform users about safe internet browsing techniques. In actuality, the best defense against malware is user education! Download programs from reliable sources only, and only visit websites that are thought to be secure. To end a window, use the keyboard shortcut Alt+F4, or utilize the Task Manager to exit out of apps, rather than clicking OK or Agree. Websites that share files should be avoided, as should the content they host. Emails containing links to potentially dangerous software downloads should be avoided
- Take into account tech that deters spyware. Use a browser, for instance, that is less vulnerable.

The following are a few typical signs of spyware:

- The default home page of the web browser has been altered.
- Every time you run a search, a certain website appears. An abundance of pop-up windows show up.
- The activity LED on the network adapter blinks a lot when the computer shouldn't be sending data.
- Antivirus and firewall software automatically shuts off.
- New favorites, icons, and programs show up.
- Windows occasionally has weird issues (slow system, strange behavior from apps, etc.).

- At random, the Java console emerges.

To troubleshoot and repair spyware-infected systems, disconnect the system from the internet or local network, uninstall the program from the Control Panel or Settings, and reboot the computer. Use antivirus software to remove any viruses infesting the system, and perform this in the recovery environment if available.

Note: When Safe Mode is insufficient, you may need to boot from a disc or USB flash drive that contains a bootable operating system or kernel (such as Knoppix) and then restart the scans.

Use anti-spyware software to quarantine and remove spyware from the computer. Be cautious with other programs, as they may modify the Registry. Only remove the malware part.

To prevent malware re-emerging on your system, check your browser's home page settings, ensure your HOSTS file isn't hijacked, and remove unwanted websites from the Trusted Sites.

Badware, including viruses, spyware, and malware, can be malicious or not. Examples include shopping toolbars that record shopping activities and purchases without user consent, or legitimate programs installing spyware or scareware without user consent. To protect against badware, users should be cautious about installing software from legitimate sources and be wary of unknown removable media before connecting it to their computer.

Rootkit Prevention and Troubleshooting

A must know!

Rootkits are unauthorized programs that can gain access to a system and act as the root or administrator user. They are detected by signature-based and heuristic-based antivirus programs. However, detecting them after execution can be difficult due to the complexity of the rootkit's modifications. Security software manufacturers attempt to detect running rootkits, but their success is unlikely. Using removable media to boot the computer makes it easier to detect the rootkit.

Rootkits can hide in MBRs, so manufacturers recommend sweeping with and antivirus software. GPT, instead of MBR, helps discourage rootkits, so use it whenever possible.

Reinstalling all software or re-image the system is the best way to combat rootkits, as it's less time-consuming and can confirm the rootkit's complete removal.

Taking Steps to Avoid and Fix Spam

The Internet needs to be conserved, as it has limited storage and bandwidth. Spam, a major gobbler of Internet resources, causes billions of dollars in fraud and productivity. Blocking spam, reporting spammers, and training users are crucial steps to conserve resources and reduce costs.

Here are a few strategies to put anti-spam security measures in place and, ideally, lessen spam:

A spam filter/filter can be purchased as software or hardware, monitoring spam activity and creating whitelists and blacklists. Network administrators should block emails with non-compliant attachments. Companies using web-hosting companies may have more spam filtering options. Clients can configure Outlook and other mail programs to have higher security against spam. Spam filters can also be installed on individual clients. Popular antivirus suites have built-in spam filtering, so ensure it is enabled.

- **Close open mail relays:** Anybody on the Internet can send email over an SMTP server, which is also known as an open mail relay. A fully open mail relay is not what the corporation that operates the SMTP server wants, even though this is what its clients want. Therefore, open mail relays ought to be secured or blocked so that only clients and users who have completed the necessary authentication can access them. SMTP open relays are another name for open mail relays.
- **Take the firm website's email address links down:** Substitute these with web forms (secure PHP forms, for instance) that allow a user to get in touch with the business without seeing any company email addresses. For any literature or advertisements, use an email account dedicated to advertising.
  Think about updating this frequently; marketers may already be tracking leads with this. To go even farther, think about adding an email service.
- **Employ blacklists and whitelists:** Blacklists are lists of email addresses or email domains that are not trusted, whereas whitelists are safe lists of trusted email addresses or whole email domains. These lists can be configured in mail client applications like Outlook as well as on mail servers and email appliances.
- **Educate your users:** Employees should use a free email address for company-related posts and screen emails carefully, using blind carbon copy (BCC) when sending emails to multiple users. Companies may enforce this

policy and monitor email habits, with some having a "safe" list where only authorized addresses can send emails to users. Training employees not to make purchases from unsolicited emails is also important.

You Can't Save Every Computer from Malware!

Sometimes, computers become infected with malware, preventing data saving. To restore data, back up data, reinstall operating system and applications, flush UEFI/BIOS, and thoroughly check for residual effects and proper hard drive performance.

An Overview of Techniques for Preventing Malware

Malware Threat	Prevention Techniques
Virus	• Launch and maintain your antivirus program. • Periodically scan the entire system. • Change the operating system. • Put in place a firewall.
Worm	• Launch and maintain your antivirus program. • Periodically scan the entire system.
Trojan Horse	• Launch and maintain your antivirus program. • Periodically scan the entire system. • Do a routine Trojan scan.
Spyware	• Launch and maintain your anti-spyware program. • Periodically scan the entire system. • Modify the browser's settings. • Think about the technology that deter spyware.
Rootkit	• Launch and maintain your antivirus program.

	• Utilize rootkit detection software.
Spam	• Put a spam filter to use. • Set up blacklists and whitelists. • Shut off open mail relays. • Educate the people you use.

CHAPTER 3:

ACRONYMS	LIST OF ABBREVIATIONS
WPA2	*Wi-Fi Protected Access 2*
AES	*Advanced Encryption Standard*
UEFI	*Unified Extensible Firmware Interface*
BIOS	*Basic Input/Output System*
IDS	*Intrusion Detection System*
HIDS	*Host-Based Intrusion Detection System*
NIDS	*Network Intrusion Detection System*
PCI DSS	*Payment Card Industry Data Security Standard*
NIST	*National Institute of Standards and Technology*
DLP	*Data Loss Prevention*
BYOD	*Bring Your Own Device*
NAS	*Network Attached Storage*
RAID	*Redundant Array of Independent Disks*
SED	*Self-Encrypting Drive*
FDE	*Full Disk Encryption*
TPM	*Trusted Platform Module*
HSM	*Hardware Security Module*
GSM	*Global System for Mobile Communications*
RFID	*Radio Frequency Identification*
CVE	*Common Vulnerabilities and Exposures*
GPS	*Global Positioning System*
SMS	*Short Message Service*
MDM	*Mobile Device Management*

Chapter 3

Computer Systems Security: Part II

Malicious software poses a significant threat to computer systems' security, affecting desktop computers, laptops, tablets, and smartphones. To eliminate these threats, proper configuration of UEFI/BIOS and security applications should be implemented on all client computers on the network, including desktop computers, laptops, mobile devices, and IoT devices. This ensures the security of all devices and systems.

Implementing Security Applications

The chapter discusses endpoint protection platforms, specifically anti-malware suites, which typically include antivirus, anti-spyware, and anti-spam components. These suites often have a built-in firewall, known as a personal firewall, and may also have an intrusion detection system (IDS) for monitoring and detecting malicious activities. There are also standalone software firewalls and HIDSs, and a pop-up blocker integrated into web browsers and add-ons. These security applications help users avoid potentially malicious websites.

Personal Software Firewalls

Quick Hits!

Personal firewalls are applications that protect an individual computer from unwanted Internet traffic through a set of rules and policies. Some personal firewalls, also known as host-based firewalls, prompt the user for permission to enable specific applications to access the Internet. Some personal firewalls have the capability to detect intrusions and block them, which is a basic form of a HIDS.

Examples of software-based personal firewalls include Windows Firewall, ZoneAlarm, PF (packet filter) and IPFW (IP Firewall), and IP cables (Internet Portable Cables). Anti-malware application suites from Symantec, McAfee, Kaspersky, and others include personal firewalls as well. This trend is expected to continue in the future.

Personal firewalls should be updated often, and auto-updates may be preferred depending on an organization's policies. As software, personal firewalls can utilize some of the computer's resources, which can cause issues with CPU power and RAM usage. Some anti-malware suites still run better than others, depending on the scenario.

Some organizations opt not to use personal firewalls on client computers and focus more on network-based firewalls and other security precautions. The choice of whether to use personal firewalls, network-based firewalls, or both can vary, but careful analysis should be performed before a decision is made. For example, an organization that accepts credit card payments over the phone and performs transactions over the Internet needs to have a personal firewall installed with the proper ports fixed to comply with financial regulations.

Personal firewalls can also be the victim of attack, as they can be affected by worms or other malware. Therefore, antivirus suites should be updated often, with daily updates being the optimal solution. A common scenario for security in small offices and home offices is to have a four-port SOHO router/firewall protecting the network and updated personal firewalls on every client computer. This combination provides two levels of protection for the average user, but larger networks usually concentrate more on network firewalls and network-based IDS.

Host-Based Intrusion Detection Systems.

Quick Hits!

Intrusion detection systems (IDSs) are used to monitor computer systems, networks, or a portion of a network and analyze data to identify incidents and attacks. There are two types of IDSs: host-based intrusion detection system (HIDS) and network intrusion detection system (NIDS). A HIDS is installed on an individual computer and can interpret encrypted traffic, but it has disadvantages such as its purchase price, resource-intensive operation, and default local storage of the HIDS object database.

An IDS can perform two main types of monitoring: statistical anomaly and signature-based. Statistical anomaly is when an IDS establishes a performance baseline based on normal network traffic evaluations and compares current network traffic activity with the baseline to detect whether it is within baseline parameters. Signature-based monitoring analyzes network traffic for predetermined attack patterns, which are stored in a database that must be updated regularly to have effect.

IDS solutions need to be accurate and updated often to avoid misidentification of legitimate traffic or attacks. Two main types of misidentification include false positive and false negative. If the IDS does not have a particular attack's signature in its database, the attack will likely not be detected, and the attack will run its course without any IDS warnings.

Some antivirus application suites have basic HIDS functionality, but true HIDS solutions are individual and separate applications that monitor log files, check for file integrity, monitor policies, detect rootkits, and alert the administrator in real time of any changes to the host. Compliance regulations, such as PCI DSS and NIST 800-53, require the type of file integrity monitoring that a HIDS can provide.

It is important to protect the HIDS database, which can be a target for attackers. It should either be encrypted, stored on read-only memory, or stored outside the system. If an IDS observes an incident, it notifies the systems or network administrator so they might quarantine and fix the problem.

In summary, intrusion detection systems (IDSs) are essential tools for monitoring computer systems, networks, and networks to detect and prevent attacks and incidents.

Pop-Up Blockers

Webmasters often advertise products and services on websites to generate revenue, charging fees to the organization that creates them. However, users quickly learn to distinguish between advertisements and non-advertisements, making advertisers constantly create new ways to advertise their products. Traditional JavaScript-based pop-up windows are no longer effective due to built-in pop-up blockers in web browsers. These ads can take a toll on user productivity and can be detrimental to the user's computer.

An attacker may create entire websites with malicious pop-ups to infect as many computers as possible. Systems administrators should try to block pop-ups using pop-up blockers, which can be enabled or added to most web browsers. However, pop-up blocking can block content that is not an advertisement but is integral to the website. Proper website functionality should be integrated directly into the actual web page, as most users ignore pop-ups or consider them malicious.

Content filtering, also known as pop-up blocking, is another level of content filtering that blocks external files that use JavaScript or images from loading into the browser. As advertisers become cleverer, proxy-based programs like Squid can filter content for multiple computers.

Advertisers have devised new tricks to get past pop-up blockers and content filters, such as flash-based pop-ups, pop-under ads, and dynamic HTML (DHTML) hover ads. Systems administrators should be prepared to update clients' web browsers and browser add-ons regularly to protect their clients' online presence.

Data Loss Prevention Systems

Cloud-based DLP solutions are essential for protecting against data breaches and misuse. These systems, which can be software or hardware-based, are designed to inspect data in use, motion, and at rest to prevent unauthorized use and data leakage. They come in three variants: endpoint DLP systems, network DLP systems, and storage DLP systems. Endpoint systems monitor data in use, such as email communications, while network DLP systems inspect data in motion. Storage DLP systems are typically installed in data centers or server rooms. Cloud-based DLP is necessary for companies with increased Bring Your Own Device (BYOD) usage and those storing data and operating infrastructure within the cloud. DLP solutions must be accurate and updated to reduce false positives and false negatives, and security administrators must determine the threat's reality.

Securing Computer Hardware and Peripherals

The article discusses the importance of securing computer hardware, BIOS, and external peripheral devices, including USB flash drives and external SATA hard drives. It suggests that a DLP solution is needed to prevent users from using USB connections and devices without proper security.

Note: Manufacturers often refer to firmware as BIOS in new computers, but UEFI-(Unified Extensible Firmware Interface) can replace or work alongside it. UEFI-only technologies are described for simplicity.

Securing the BIOS

The BIOS (basic input/output system) is a crucial component of a computer system that can be vulnerable to malicious attacks and acts as a gateway to the rest of the system. To protect it, follow these steps:

1. Flash the BIOS: Flashing the BIOS to the latest version can prevent exploits and errors, as well as protect against electromagnetic interference and electromagnetic pulses. Newer motherboards issue at least one new BIOS version within the first six months of the motherboard's release.
2. Use a BIOS password: The supervisor password blocks unwanted persons from accessing the BIOS. Some organizations opt to use locking cables or similar devices to deter unauthorized access.
3. Configure the BIOS boot order: Set up the BIOS to reduce the risk of infiltration. For example, change the BIOS boot order to look for a hard drive first.

4. Disable external ports and devices: Disable removable media if a company policy requires it.
5. Enable the secure boot option: UEFI 2.3.1 and higher offer the secure boot option, which prevents unsigned or improperly signed device drivers and OS loaders. Ensuring secure boot is working properly is crucial, and the measured boot option is used to verify the boot process's integrity.

Securing Storage Devices

Storage devices, often with moving parts or removable media, pose security risks and can be difficult to track. Solutions include physical security, encryption, and policies. This section covers removable storage devices, network attached storage (NAS), whole disk encryption, and hardware security modules (HSMs).

Removable Storage

Removable storage, including optical discs, USB devices, eSATA devices, and laptop disks, can be controlled by network administrators through the BIOS and operating system policies. Companies often block all removable media except for necessary devices, which are approved on a case-by-case basis. Users should be trained on proper usage and not allowed to take data home. USB devices can be vulnerable to attacks, as they can carry large amounts of data and act as delivery mechanisms for malware. Organizations must decide whether to allow USB devices and implement security procedures. Data on USB devices can be encrypted using software programs or purchased from secure brands like IronKey. A security administrator should implement removable media controls, including USB lockdown, limited use, malware scanning, data encryption, monitoring, and auditing. Reuse and disposal of media should be managed carefully.

Network Attached Storage

Network attached storage (NAS) is a storage device connected to an Ethernet network, typically containing two hard drives for RAID 1 mirroring. Advanced NAS devices can house up to 32 drives and terabytes of data. RAID arrays aim for high availability with 99.9% average uptime. Network security is crucial when implementing NAS, including data encryption, authentication, and secure logging.

Whole Disk Encryption

Quick Hits!

Encryption is crucial in computer security, ensuring data is only accessible to authorized individuals. To encrypt an entire hard drive, a self-encrypting drive (SED) or full disk encryption (FDE) software is required. Microsoft's BitLocker is a popular FDE software, available in elite editions of Windows. Requirements include a trusted platform module (TPM), an external USB key, and a hard drive with two volumes. BitLocker is based on the Advanced Encryption Standard (AES) and can use 128-bit and 256-bit keys. BitLocker security settings can be accessed in the appropriate Windows version. Double encryption, such as using EFS to encrypt individual files, can be a successful technique for securing files.

Hardware Security Modules

Hardware security modules (HSMs) are physical devices used for secure crypto processing, including login/authentication processes, digital signings, and payment security systems. They are faster than software encryption and can be found in adapter card form, USB-connected devices, or network-attached devices. HSMs are tamper-proof and can work independently in high-availability clustered environments. However, they often require management software installation, leading to slower deployment in some organizations. HSMs are often involved in generating, storing, and archiving encrypted key pairs.

Securing Wireless Peripherals

Wireless devices, like keyboards and mice, can be easily intercepted due to their spread spectrum. To protect against this, some manufacturers offer encrypted wireless device solutions, such as AES. Wireless displays, cameras, SD/microSD cards, external storage devices, printers, and multi-function devices should use WPA2 and AES for connections. Proper authentication and strong encryption are essential for wireless peripherals, and it's important to regularly update wireless security schemes.

Securing Mobile Devices

Mobile devices, including smartphones and tablets, can be targets for attacks, potentially causing service abuse or larger-scale attacks. To protect users, they should be cautious about sharing their phone numbers, avoid unsolicited emails or text messages, and educate them about the issues with downloadable software. Use a complex password when necessary and update mobile operating system software like desktop computer software. Encrypt data in various ways, and some organizations have policies specifying how data will be encrypted. The National Cyber Awareness System (NCAS) and U.S. Computer Emergency Readiness Team (US-CERT) provide general tips for encrypting data. Mobile devices, including laptops, media players, e-readers, gaming devices, cell phones, and fitness devices, are particularly vulnerable to attacks.

Malware

Malware is present in all operating systems, including Windows, and Android has been a significant target for malware. Viruses, worms, rootkits, and other malware are common in Android OS and iOS. Mobile operating systems should be updated to the latest version and antivirus software can be used. Regularly updating security programs and avoiding sharing personal or organizational information on social media networks are crucial.

Botnet Activity

Mobile devices can be part of botnets, similar to desktop computers, and can launch DDoS attacks, click fraud, or scams. To prevent botnet involvement, users should download legitimate apps, avoid rooting or jailbreaking devices, and use custom firmware. Over-the-air firmware updates may not work anymore, and if a device appears to be part of a botnet, it should be wiped. Similar to malware prevention, mobile devices should be wiped if they appear to be part of a botnet.

SIM Cloning and Carrier Unlocking

SIM cloning is a method where an attacker gains access to all phone data by using the same service. V1 SIM cards have a weak algorithm, while V2 cards are harder to clone due to a stronger chip. Users and administrators should be aware of the SIM card version and update it if needed. Unlocking a smartphone from its carrier is also a risk, as it takes it off the grid and makes it difficult to track and manage.

Wireless Attacks

Wireless services like 4G, 3G, GSM, Wi-Fi, infrared, RFID, and Bluetooth pose security risks. To minimize these risks, turn off the service when not in use, use airplane mode, or turn off the mobile device altogether. Bluetooth is particularly vulnerable to virus attacks, bluejacking, and bluesnarfi ng. To prevent bluejacking, set the device to "undiscoverable" or turn off Bluetooth altogether. Discourage bluesnarfi ng by using a unique pairing key or setting devices to "undiscoverable" after legitimate setup. Wi-Fi also has vulnerabilities, and security administrators must monitor current CVEs and available updates and patches. This applies to smartphones and IoT devices with built-in Wi-Fi connections.

Theft

Mobile devices are missing or stolen every minute, with theft being the worst attack. To protect against data loss and recover from theft, organizations should use data encryption, GPS tracking, remote lockout programs, remote wipe programs, screen locks, complex passwords, and user training. These measures help prevent thievery and protect sensitive data. Additionally, implementing remote lockout programs and sanitizing phones remotely can help prevent data loss. User training can also help users report lost or stolen devices to law enforcement.

Application Security

Mobile applications are crucial for data security, and it is essential to ensure proper user interaction and encryption. Encryption is a key method to protect data and prevent potential attackers from interfering with applications. Whole device encryption, which encrypts internal memory and removable cards, is recommended by security administrators. Data in transit, which occurs between a client and a server, is another important aspect of mobile application security. Most mobile devices communicate with servers, such as web browsers, email clients, contacts databases, or apps that work independently of a browser but connect to a server.

Servers play a crucial role in key management, which involves creating, storing, using, and reviving encryption keys. Organizations often purchase master key algorithms from third-party companies like VeriSign, who inform the organization if a key has been compromised and needs to be revoked. Credential management is also important for security administrators.

Transitive trust, where users access data on other networks, can be insecure and allow attackers to access other networks. Organizations typically prefer non-transitive trust, where users need to be authenticated separately to each network and have access to only their own applications. Application whitelisting, which restricts users to only approved applications, is often used to increase application security. This can be done within a computer policy or through a mobile device management system.

Messaging apps are particularly vulnerable to malware, and unwary users of mobile devices are especially susceptible to Trojans and phishing via SMS texts. To prevent this, mobile anti-malware should be installed, controlled from a mobile device management solution, or blocked messaging apps altogether or using company-approved apps.

Mobile payment methods in organizations can be vulnerable to vulnerabilities such as weak passwords, user error, and phishing. Users should be educated on using payment apps discreetly and avoiding loss or theft. Security administrators should consider using external readers for transactions and teaching users to keep them separate from the device. Geotagging, which adds location-specific information, can pose security threats, especially for executives with confidential information. BYOD concerns also arise from mobile devices.

BYOD Concerns

In 2011, organizations began allowing employees to bring their own mobile devices into work and connect them to the organization's network. This "bring your own device" concept has become popular but can be challenging for security administrators and users. To successfully implement BYOD, storage segmentation is crucial, with clear separation of organizational and personal information, applications, and content. Mobile device management (MDM) platforms can help manage tasks such as application management, content management, and patch management. MDM software also allows for easier access control, inventory control, and enabling unused features on multiple devices.

However, user acceptance of BYOD is mixed, with some liking the idea of using their own devices and others believing it is just a way to move computing costs from the company to the user. Legal concerns, such as privacy, employee misconduct, and fraud, also arise. To address these concerns, organizations should establish firm policies about acceptable use, data ownership, and support ownership. Third-party providers offer embedded certificate authority, posture monitoring, and certificate-based security to help secure mobile devices in a BYOD enterprise environment.

CHAPTER 4:

ACRONYMS	LIST OF ABBREVIATION
VM	*Virtual machines.*
VPN	*Virtual private network*
TOS	*Trusted Operating System*
OS	*Operating System*
SCCM	*System Center Configuration Manager*
GPO	*Group Policy Object*
NTFS	*(New Technology File System).*
NTFS	*New Technology File System*
KB	*Knowledge Base*
G PL	*GNU General Public License*
MDM	*Mobile Device Management*
BIOS	*Basic Input/Output System*
UEFI	*Unified Extensible Firmware Interface*
VPN	*Virtual Private Network*
VDI	*Virtual Desktop Infrastructure*
VLAN	*Virtual Local Area Network*
VM	*Virtual Machine*
VMM	*Virtual Machine Manager*
IaaS	*Infrastructure as a Service*
CVE	*Common Vulnerabilities and Exposures*

Chapter 4

Hardening Operating Systems and Security Best Practices

A non-hardened server operating system can be compromised quickly, depending on the organization's size and popularity. Regular updates to servers, workstations, routers, and switches are crucial for security. Employing group policies and baselining hardens systems, making them tough enough to withstand the pounding of modern technology and society.

Virtual operating systems offer high security, portability, and ease of use, but are resource-intensive. Balancing virtualization with organizational resources is crucial. These systems require maintenance and hardening, ensuring a secure environment.

Effective virtualization and intelligent update plans can enhance security and longevity of operating systems and their relationships.

Hardening Operating Systems

Operating systems installed out-of-the-box are inherently insecure due to initial code issues, product age, and basic rules. Despite improvements, new applications and technologies offer new security implications. It's crucial to protect the system after installation to ensure a secure experience.

Hardening an OS involves securely configuring it, updating it, creating rules, and removing unnecessary applications to minimize threats and mitigate potential risks. The text emphasizes that a "bulletproof" system is not possible, as no system can be 100% secure. Instead, it suggests methods to reduce current and future risk to an acceptable level, although it is impossible to achieve zero security.

This section outlines methods to harden the OS using patches, patch management, hotfixes, group policies, security templates, and configuration baselines. It also discusses securing the system and hard drives, analyzing the system to identify unnecessary applications and services.

Removing Unnecessary Applications and Services

Organizations often use least functionality to limit the use of resources and vulnerabilities in their operating systems. This approach involves limiting the use of computers and information systems to only provide essential functions, thereby reducing potential security risks.

A proactive mindset can protect systems from threats from insecure applications and services, such as instant messaging programs. These programs can be dangerous due to their backdoors and lack of productivity. Unless required by tech support, they should be discouraged or disallowed by rules and policies. If necessary, remove all traces of the application. Avoid remote control programs, such as Remote Desktop Connection, which uses inbound port 3389, a known security threat. Use a different port if necessary, disable associated services, and ensure inbound ports are closed and secured. Disable any shares created by the application and remove all instances or re-image the computer if necessary.

To conserve hard drive space, remove unnecessary programs like Camtasia, Pro Tools, or Premiere Pro from the Notification Area in Windows. These programs use processor/RAM resources and can be removed by right-clicking and accessing their properties or using a control program like Task Manager or Misconfiguration.

The app's potential internet communication for updates raises resource and security concerns, requiring its removal if unused and only installing necessary software in the future.

For a large network with 1,000 computers, centralized management systems like Microsoft's SCCM-(Microsoft System Center Configuration Manager) and mobile device management suites allow security administrators to manage software, configurations, and policies from a local workstation, eliminating the need for local removal of applications.

To protect end-user computers from malicious applications, consider using application whitelisting and blacklisting. Whitelisting allows only certain applications to run on client computers, such as Microsoft Word and a secure web browser, while denying other applications. This eliminates hacking risks but can cause productivity issues. If an end user needs another application, exceptions must be made, which can take time and management permission. Blacklisting, on the other hand, disallows individual applications across the network, making it more efficient for end users who frequently use multiple applications. Both whitelisting and blacklisting are typically performed from centralized management systems and policies.

Removing applications can reduce attack surface and improve computer performance, but underlying services, used by applications and OS, can also be a security burden and resource burden.

The Remote Desktop Services service is currently running. To stop and disable it, right-click on it, select Properties, click Stop, apply, and change the Startup Type dropdown menu to the Disabled option.

Disabling unnecessary services, particularly on Windows servers, can reduce the risk of unwanted access and reduce the amount of resources used, especially as they run more services and are a common target, thus reducing the size of the attack surface.

The Windows Command Prompt allows for the start and stop of services using the net start and net stop commands, as well as the sc command.

Note: You will need to run the Command Prompt in elevated mode (as an administrator) to execute these commands.

We stopped and started the Windows Firewall service using net stop and net start commands. The service stopped immediately, and we used the sc command to disable it, ensuring it wouldn't start again when the system was restarted.

The command-line in Linux allows users to start, stop, and restart services. To display a list of services and their status, use the command-line first. For example, in Ubuntu, you can use service -—status-all. For upstart jobs, use initial list. The sc query command can be used to find the service name.

Note: If you are not signed in as an administrator (and even if you are), you must put sudo before a command and be prepared to supply an administrator password. Prepare to use sudo frequently.

Services can be stopped in the Linux command-line in a few ways:

1. Enter the following syntax:
   /etc/init.d/service>
2. Stop, where service > is the name of the service. For instance, if you have become an Apache web server, you would type: stop /etc/init.d/apache2
3. Services can also be initiated and restarted by substituting start or restart for stop.
4. In select versions, type the following syntax: stop service service>

Some services need a distinct set of syntax. Telnet, for example, may be removed under Red Hat by running check configuration telnet off. Check the MAN pages for your specific version of Linux, either on the command line or online, for exact syntax and any earlier instructions that must be executed. Alternatively, use a generic Linux online MAN page, such as http://linux.die.net/man/1/ telnet.

Services in macOS/OS X Server may be terminated from the command line using the following syntax: sudo serveradmin service> stop

This may not function on the macOS or OS X client. In 10.9 Mavericks, for example, you would simply terminate processes using the Activity Monitor or the kill command in the Terminal.

Note: When the service cannot be halted, it is occasionally required in an operating system to terminate the underlying process. In Windows, use the Task Manager or the task kill command to terminate processes.

The task kill command can be used in combination with either the task's executable name or the process ID (PID) number. The task list command may be used to find numbers associated with processes. To kill an individual ID that is presently operating on Linux, use the syntax ps aux | less. (If required, use Ctrl+Z to exit instructions like this.) The equivalent on mobile devices would be a forcible quit.

Let us not overlook mobile gadgets. You may need to force stop or delete programs from Android or iOS. Prepare yourself.

Summary of Ways to Stop Services

Operating System	Procedure to Stop Service
Windows	Access services.msc from the Run prompt. Use the net stop <servicename> command in the Command Prompt. Use the sc stop <servicename> command in the Command Prompt.
Linux	Use the syntax /etc/init.d/<servicename> stop. Use the syntax service <servicename> stop (in select versions). Use the syntax chkconfig <servicename> off (in select versions).

macOS/OS X	Use the kill command to end processes. Also works in Linux. In Windows, this is the task kill command.

Windows Update, Patches, and Hotfixes

Very Important!

A Trusted Operating System (TOS) meets government requirements for multilevel security and updates, with examples like Windows 7, OS X 10.6, FreeBSD, and Red Hat Enterprise Server. Manufacturers must have strong policies.

Regular updates are essential for operating systems, even without being a TOS. Microsoft recognizes OS vulnerabilities and potential exploits, releasing patches to enhance performance and protect the system.

To update a system, first determine the version number, build number, and patch level. In Windows, use the System Information tool, winver command, system info command in Command Prompt.

To ensure system safety, check your organization's policies and test updates on a clean, dedicated network before going live with an update.

Windows Update manages system updates, accessible in Control Panel, Run prompt, or Settings section, and can be accessed in Windows 10 through MS-settings: windows update.

Updates are divided into different categories:

Very Important!

- **Security update**: A widely released fix for a product-specific security-related vulnerability, rated based on severity in the Microsoft Security Bulletin.

- **Critical update:** has been released to address a specific, non-security-related bug in a widely released fix.
- **Service pack:** Service packs are a cumulative set of hotfixes, security updates, critical fixes, and additional fixes for internal product issues. They may contain customer-requested design changes or features. Windows 7 and Windows Server 2008 R2 are the last Microsoft operating systems to use service packs.
- **Windows update:** The recommended update addresses a noncritical issue that some users may encounter and adds features and updates to the features bundled with Windows.
- **Driver update:** A security administrator must ensure the compatibility of installed device drivers with a system by avoiding driver shimming and driver refactoring. Shimming intercepts API calls, while refactoring restructures driver code. Proper signing and testing on a closed system are crucial for preventing future malicious code issues.

Windows updates come in various options, including automatic install, defer, download with option to install, manual check, and never check. Administrators should specify what, when, and to which computers to avoid issues like patch level mismatch and network usage.

Your organization can disable Windows Update by either turning it off entirely or specifying its stop and disablement. This can be done through the Services console window or Command Prompt.

Patches and Hotfixes

Quick Hits!

Patches and hotfixes are often used interchangeably, with Windows updates consisting of hotfixes. Originally defined as a single patch to an OS or application installed live, hotfixes can be single patches to individual applications or affect the entire system. They can be identified with the letters KB followed by seven numbers.

Patches are small fixes for security issues on computers, often referred to as software updates or service packs. For instance, Microsoft released a patch for older Mac versions to prevent Trojan attacks, disabling remote access.

To install a patch, check if it was part of a group update, such as a macOS or OSX patch for iTunes. Determine the OS version, as it might already include the patch, by clicking the Apple menu.

Note: All systems, including Windows, Mac, Linux, UNIX, Android, iOS, hardware appliances, kiosks, and automotive computers, require patching at some point.

Patch design can cause software regression, so patch management is crucial to ensure accurate system behavior and interactions, as patch design can be flawed.

Commonly Asked!

Patch Management

Patch management involves planning, testing, implementing, and auditing patches to ensure proper security and prevent random updates. Different companies may have different strategies, but each should include these four steps.

- **Planning:** A plan should be established to determine the necessity and compatibility of a patch, using tools like Microsoft Baseline Security Analyzer. If necessary, the plan should include testing in a clean network, implementation timeline, and post-installation check-in procedures.

- **Testing:** To ensure successful patch deployment, it's essential to test it on a single system or small group of systems, avoiding regular network usage by "civilians" or regular users. This can save resources and reduce the risk of failure.
- **Implementing:** Successful patch testing should be performed on all necessary systems, often in the evening or over the weekend, using software like Microsoft's SCCM and third-party patch management tools.
- **Auditing:** After implementation, audit systems to ensure patch takes hold and detect changes or failures, using SCCM and third-party tools to ensure proper patch implementation.

Linux and Mac-based programs and services help manage patching and auditing. Red Hat offers services for system admins, while third-party services are available for GPL Linux users. Mobile Device Management (MDM) platforms benefit networks with many devices. However, patch management can be overwhelming for some organizations.

Group Policies, Security Templates, and Configuration Baselines

Hardening an operating system involves removing applications, disabling services, patching, hot-fixing, and installing service packs. Administrative privileges should be used sparingly, and policies should be in place to enforce organizational rules. Secure configurations and baselines should be initiated to measure server and network activity.

To access the Group Policy in Windows, navigate to the Run prompt and type gpedit.msc, which will open the Local Group Policy Editor Console window.

This guide focuses on computer security settings, accessible through Local Computer Policy > Computer Configuration > Windows Settings > Security Settings. It covers password policies, account lockout policies, and public key policies. The Local Group Policy Editor governs specific machines and users, used by Windows Server domain controllers with Active Directory loaded.

Security templates, groups of policies loaded in one procedure, are commonly used in corporate environments. They can be installed by right-clicking Security Settings and selecting Import Policy, especially important on Windows Server computers.

Windows Server has three main security templates: defl tbase.inf, defl tsv.inf, and defl tdc.inf. These templates are stored in %systemroot%\inf. To open a policy, select it and click Open. These templates are typically applied to domain controllers but can be used for other systems.

Windows Server allows policy modification and template addition directly from Server Manager > Security Configuration Wizard, saving templates as.xml files instead of.inf files.

Group Policies, loaded with various GPOs, implement OS hardening and host-based security for workstations by confining as many as possible.

Baselining is a process of consistently measuring changes in networking, hardware, and software. It involves selecting a specific data point and measuring it consistently for a period of time. This baseline can help determine the standard load and server requirements and is crucial for testing equipment and monitoring installed devices. However, baseline deviations are often overlooked by security administrators, especially when dealing with real-time operating systems (RTOS). Baselining is especially important when dealing with systems that require near 100% uptime and lightning-fast response with zero latency.

Hardening File Systems and Hard Drives

This chapter provides tips on hardening a hard drive and its file system. On Microsoft computers, NTFS-(New Technology File System) is the best option due to its security, logging, encryption, and larger partition sizes. FAT32 and NTFS support the same amount of file formats, making NTFS the best choice. Converting a volume to NTFS is easy.

Convert volume /FS: NTFS

For example, if I want to convert a USB flash drive named M: to NTFS, the syntax would be **Convert M: /FS: NTFS**

The convert command offers additional options, such as converting */?* in the Command Prompt. NTFS provides file-level security and tracks permissions within access control lists. Most systems use NTFS, but flash-based and removable media can be identified using the chkdsk command or the Properties command.

System files and folders are typically hidden to protect a Windows system. To permanently hide them, go to File Explorer Options, select the View tab, and select Don't Show Hidden Files, Folders, or Drives. To hide protected files, select Hide Protected Operating System Files. Additionally, turn off file sharing.

Hard disks can fail, but it's not a matter of if; it's a matter of when. By maintaining and hardening them with various utilities, we can prevent that dark day.

1. **Remove temporary files:** Temporary and older files can clog hard disks, affecting performance and security. Disk Cleanup or similar programs are recommended, with policies for daily or logoff.
2. **Periodically check system files:** Regularly verifying the integrity of operating system files is recommended, and this can be done through various methods.
3. The **chk-dsk** command in Windows examines disks and fixes errors. The SFC (System File Checker) command checks and replaces protected system files, addressing OS and application problems. In Linux, the fsck command checks and repairs a Linux file system. A derivative, e2fsck, is used for Linux ext2fs. Open source data integrity tools like Tripwire can be downloaded for Linux.
4. **Defragment drives:** Hard drive fragmentation can cause server failure, affecting timely request service. Defragmenting can be done using Microsoft's Disk Defragmenter, command-line defragment, or third-party programs.
5. **Back up data:** Backing up data is crucial for a company, including individual files or entire systems. Microsoft domain controllers' Active Directory databases are particularly vulnerable to attack, so it's essential to back up the System State for future recovery.

6. **Use restoration techniques:** Restore points are essential for Windows servers and workstations to fix issues caused by defective hardware or software. The System Restore utility can fix these by reverting back to an earlier time. Restore points can be created manually or automatically by the OS. In macOS and OS X, the Time Machine utility works similarly. Linux users can back up the ~/home directory to a separate partition.
7. **Consider whole disk encryption:** Whole disk encryption is a method that secures the contents of a drive, making it harder for attackers to access and interpret its data.

The author recommends separating the operating system (OS) from the data on a separate hard drive to protect against malware and data loss. The OS-inhabited hard drive can be wiped and reinstalled without data loss, and settings should be backed up. Keeping a well-maintained computer effectively hardens the OS.

Managing a Well-Maintained Computer

- **Use a surge protector or UPS:** If you are concerned about power outages, connect the computer and other devices to a surge protector or, better yet, a UPS.

- **BIOS and/or UEFI updates:** It is not always essential to flash the BIOS; check the manufacturer's website for your motherboard to determine whether an update is required.

- **Update the operating system:** Windows involves hotfixes, updates, and notification of new updates, while Linux and macOS/OS X require upgrading and applying individual fixes.

- **Update anti-malware:** This involves ensuring that the anti-malware (antivirus and anti-spyware) software has a current license and that updates are enabled and that the program is constantly scanning the system.

- **Update the firewall:** Make sure you have a firewall installed and enabled, and then update it. If it is Windows Firewall, upgrades should be performed automatically via Windows Update. If you have a SOHO router with a built-in firewall or another firewall device, you must upgrade the ROM by obtaining the most recent image from the manufacturer's website.

- **Maintaining the disks**: This entails running a disk cleanup tool on a regular basis and checking to see whether the hard drive needs to be defragmented once a week to once a month, depending on usage. It also includes making restoration points, doing computer backups, and utilizing third-party backup or disk imaging tools.

- **Optional) Create an image of the system:** After completing OS hardening and configurations, consider creating an image of the system. This allows for quick and efficient reinstallation in case of system failure or compromise. Imaging software can be expensive, but it's worth it if you're concerned about the time it would take to restore the system. Standardized images in organizations ensure a snapshot of the system, with embedded security.

Virtualization creates a virtual entity, typically an OS-hosted machine, with various types, purposes, and applications.

Types of Virtualization and Their Purposes

Cloud Engineer Quick Hits!

Virtualization, including network, storage, hardware, and software, is a crucial aspect of the CompTIA Security+ exam. Virtual machine software allows for the simultaneous running of multiple operating systems from a single PC, benefiting programmers, developers, and systems administrators. Security researchers use virtual machines to execute malware without risking the actual OS and hardware.

Virtual machines consume resources like processor and RAM, and can cause hardware compatibility issues. If the physical computer fails, the virtual OS will automatically shut down, affecting all other virtual computers. Additionally, technicians often forget to update virtual machines with the latest service packs and patches. Organizations often define standardized virtual images for servers, which ensures mandated security updates and service packs. This image can be used in a virtual environment or copied to a physical hard drive, saving time and effort in the long run.

Virtual machines are classified into two types:

- **System Virtual Machine:** A comprehensive platform designed to replace a whole computer, allowing you to run an entire operating system virtually.

- **Process Virtual Machine:** A computer that is designed to execute a single application, such as a virtual web browser.

Virtual machines (VMs) are virtual machines that cannot cross software boundaries, providing a secure platform for testing and malware analysis. They are often used interchangeably with emulators, simulations, and virtualization. Virtual appliances are virtual machine images designed to run on virtualization platforms, such as virtual browsers. Companies like VMware and Microsoft refer to these images as virtual machines. Examples of virtualization include virtual private networks (VPN), virtual desktop infrastructure (VDI), and virtual local area networks (VLAN). Virtualization allows for isolation of network services and roles, preventing the spread of adverse effects.

Hypervisor

Virtual machine software hosts multiple virtual machines (VMs) to facilitate efficient communication and communication between them. A hypervisor, also known as a virtual machine manager (VMM), enables simultaneous running of multiple VMs on a single computer. This is due to confusion concerning the two different types of hypervisors:

- **Type 1–Native:** A bare metal hypervisor, like VMware vCenter, vSphere, Citrix XenServer, and Microsoft Hyper-V, runs directly on the host computer's hardware, either as a standalone product or as a role within Windows Server 2008 (R2) or higher.

- **Type 2–Hosted:** Hypervisor runs on top of operating system, while guest operating systems run within it, resulting in less efficient performance, like VirtualBox, Windows Virtual PC, Hyper-V, and VMware Workstation.

Type 1 hypervisors are faster, more efficient, and elastic, making them popular among web-hosting companies and cloud computing providers like IaaS. They can adjust resource supply as needed, making them more efficient than Type 2 hypervisors. However, Type 1 hypervisors have more stringent hardware and software requirements, making them less suitable for developing and testing environments. Application containerization is a type of virtualization that enables organizations to deploy and run distributed applications without launching a whole virtual machine, offering efficiency and portability but potentially exposing security threats.

Securing Virtual Machines

The security of a virtual machine operating system is similar to that of a physical machine OS. It's crucial to update VMs to the latest service pack and ensure patch compatibility. VMs should have the latest AV definitions, strong passwords, and personal firewall. Addressing these issues can prevent system downtime.

Ensure you use the latest virtual machine software, update to the latest patch, and confirm any security settings before creating virtual machines, focusing on standardized imaging.

Monitor network shares between virtual machines and physical machines, as active network shares can create a directory traversal vulnerability, allowing malware to spread between systems if properly separated.

Use network shares between virtual machines (VMs) carefully, document their connections, and regularly review them. Be cautious with VMs using bridged connections instead of NAT, as this can allow malware and attacks to bypass the bridge.

Virtual machine (VM) isolation can fail due to vulnerabilities like buffer overflows, remote code execution, and directory traversals. Security administrators should stay updated on CVEs and close unnecessary network shares and connections. If a virtual host is connected to a NAS or SAN, it's advisable to separate the storage devices off the LAN, secure it with a strong firewall, and prohibit unprotected file transfer protocols.

Disable unnecessary hardware within the virtual machine, enable removable media, and disable it after finishing. Disable devices from the virtual machine software and ensure boot priority in the virtual BIOS, ensuring the hard drive is booted first.

Virtual Machines (VMs) can be compromised, potentially leading to denial-of-service attacks. To protect against this, limit VM usage and periodically monitor their usage. Be cautious of monitoring from the main host, enabling only authorized users and disabling when not needed. The more virtual machines (VMs) an organization has, the more resource conservation is crucial. Over time, VMs become harder to manage, leading to virtualization sprawl. To reduce this, administrators can create an organized library of VM images and use VMLM tools.

Protect the raw virtual disk file by setting permissions, implementing logging and auditing if supported by your virtual machine software, and making a snapshot or checkpoint. Encrypt the file, digitally sign it, and validate the signature before usage to prevent potential physical disk disaster.

CHAPTER 5:

ACRONYMS	LIST OF ABBREVIATION
SDLC	*Software Development Lifecycle*
DHCP	*Dynamic Host Configuration Protocol*
SQL	*Structured Query Language*
HTTP	*Hypertext Transfer Protocol*
SSL	*Secure Sockets Layer*
SDLC	*Software Development Life Cycle*
UAC	*User Account Control*
PHP	*Hypertext Preprocessor*
SQL	*Structured Query Language*
MySQL	*My Structured Query Language*

Chapter 5

Application Security

Key Topic!

Browser security is crucial for any security administrator, as it is an inherently insecure software. While browsers are becoming more secure, new malware presents challenges. Most browsers have built-in options for security, and third-party programs can help. However, users also use office applications frequently, such as the command-line and back office applications. It is essential to secure any application used or developed on your network, as it takes only one vulnerable application to compromise network security.

Securing the Browser

The debate over the best web browser is significant, but it's important to plan for performance and security from the start. As a security expert, I spend time securing various browsers, but each has advantages depending on the environment. To ensure browser security, it's recommended not to use the latest versions, as they may introduce issues and backdoors. For average users and fast-paced organizations, a rock-solid browser is essential. Companies often implement this policy to avoid unnecessary installation costs and compatibility issues.

Note: Some browsers automatically update, preventing vulnerabilities, but this may not align with your organization's policies. Disabling this feature in settings or computer policies is necessary.

When choosing a browser, consider the type of computer running the browser, such as Macs running Safari, Linux running Firefox, Windows running Edge and Internet Explorer (IE), Firefox and Google Chrome, and whether you will be centrally managing multiple client computers' browsers. Consider the browser companies' vulnerabilities and whether specific websites recommend or require a specific browser. Most browsers work well enough, but one might be better for specific purposes. Remember that web browsers were not originally intended to do all the things they now do, and this trend is likely to continue with more new versions, patching, and secure code.

General Browser Security Procedures

Key Topic!

First, some general procedures should be implemented regardless of the browser your organization Uses.

These concepts can be applied to desktop browsers as well as mobile browsers.

- Implement policies.
- Train your users.
- Use a proxy and content filter. Secure against malicious code.

Implement Policies.

A policy can be handwritten, implemented at the browser, computer operating system, or centrally. It can manage add-ons, disallow access to malicious websites, have Flash content, or use excessive bandwidth. To modify security settings, open the Local Group Policy Editor, navigate to Administrative Templates, and select Internet Explorer or Microsoft Edge.

The image shows the folder containing Microsoft Edge policy settings and Internet Explorer open at the top of the window. There are numerous settings for Internet Explorer, and you can modify Internet Explorer Maintenance Security by navigating to User Configuration > Windows Settings > Internet Explorer Maintenance > Security. Most security features can be configured directly within the browser if not in the OS.

To update Internet Explorer security policies on multiple computers, use a template or a domain controller to centralize the changes. This allows for more in-depth security to be ensured and deployed to IE browsers within multiple computers. An example of managing IE policies using a domain controller is shown in a Windows Server setting up an organizational unit (OU) named Marketing and creating a Group Policy object named Marketing-Policy. This policy can be accessed by navigating to Computer Configuration > Policies > Administrative Templates > Windows Components > Internet Explorer, enabling the provision of trusted and non-trusted sites, zones, and advanced security features for all computers in the OU.

Train Your Users

User training is crucial for understanding websites, search engine usage, and handling pop-ups. Organizations can provide onsite training classes, webinars, and screencasts, or write engaging web articles. For example, users should learn to press Alt+F4 to close pop-up windows instead of clicking No or an X. If this doesn't work, applications can be closed by ending the task or process, force stopping, or force quitting.

To ensure web communications are secure, users should check if the website uses proper encryption certificates. Some browsers use a padlock icon in the locked position to show security. To find the specific security employed by the website, click or right-click on the padlock icon and select More Information or Connection. This will show the encryption type and certificate being used.

Use a Proxy and Content Filter

HTTP proxies act as a mediator between network clients and the Internet, reducing the number of requests needed to forward to the corresponding web server. This saves time, efficient bandwidth use, and secures client connections. Content filters can filter out potentially malicious websites, such as peer-to-peer (P2P) websites. To connect to a proxy server, the client's web browser must be trusted to connect. This can be done automatically or statically. Alternatively, it can be configured within an organizational unit's Group Policy object on the domain controller. However, any networking connection can be used for evil purposes, as malicious individuals can use malware to write their own proxy connection to the client operating system, redirecting users to potentially harmful websites. As a security administrator, it is crucial to enable legitimate proxy connections and disable illegitimate connections used by attackers.

Secure Against Malicious Code

The chapter discusses the need for increased security measures in various software applications, including ActiveX controls, Java, JavaScript, Flash media, and phishing, depending on the company's policies and procedures.

o Web Browser Concerns and Security Methods

To enhance web browser security, companies should find a balance between functionality and security. However, in enterprise or mission-critical environments, a greater focus on security is necessary, as more secure browsers can reduce functionality.

- ### Basic Browser Security

To ensure web browser security, update the browser if company policy permits it and use the patch management strategy. Install pop-up blocking and ad-blocking solutions, such as antivirus suites and third-party add-ons. Newer web browsers may block some pop-ups on their own. Consider security zones, setting security levels for the Internet and intranet zones, trusted and restricted sites, and custom levels like disabled ActiveX controls and Java applet scripting. Avoid jailbreaking or rooting mobile devices to keep the browser secure and the entire system. This will make it easier for attackers to attack the web browser and subjugate the system.

- ### Cookies

Cookies are text files placed on a client computer that store information about the user, including browsing habits and user credentials. Persistent cookies are used to prevent users from logging in to a website repeatedly. Users can adjust cookie settings to accept or deny all cookies, with high security settings typically blocking cookies that save sensitive information or do not have a privacy policy. Users can also override automatic cookie handling by confining the prompt option, which prompts the user when a website attempts to create a cookie. However, there is a learning curve for users when it comes to accepting cookies.

Spyware uses tracking cookies to collect information about a user's activities. Cookies can also be targeted for attacks, such as session hijacking, where an attacker manipulates a client computer into executing code considered trusted as if it came from the server the client was connected to. This can lead to the acquisition of the session cookie and other security threats.

- **LSOs**

Adobe Flash-based websites store data on users' computers, particularly for Flash games, raising privacy concerns. Local Settings Manager (LSOs) can be disabled in most operating systems, or completely removed with third-party software or by accessing the user's profile folder in Windows. The typical path is,C:\Users\[YourProfile]\AppData\Roaming\Macromedia\Flash Player\#SharedObjects\[variable folder name].

- **Add-ons**

Add-on programs for your browser can be enabled and disabled through settings or the add-ons management utility. However, be cautious as they can pose a risk to the browser, system, or network. ActiveX controls, similar to Java applets, can execute programs on Internet Explorer and Windows operating systems. These can contain viruses, spyware, or other malicious add-ons, particularly Flash scripts. To disable undesirable scripts, disable them from advanced settings or create a custom security level or zone. If script technology cannot be disabled within the browser, consider using a different browser or a content filtering solution. Depending on the add-on and situation, other solutions include updating Flash and upgrading the browser.

o Advanced Browser Security

To secure a web browser, locate the settings or options section in the upper-right corner of the web browser and select the advanced tab of the Internet Properties dialog box. Automatically flush temporary browser files from a system every day is recommended, as they can contain personal identifiable information (PII). Remote users should be trained to delete temporary files, cookies, and passwords when using computers on the road. Most companies discourage saving passwords by the browser, but some organizations make it a policy to disable that option. If saved passwords are necessary, enter a master password and use a password quality meter to determine its strength.

A company might use the advanced settings of a browser to specify the minimum version of SSL or TLS for secure connections. Some organizations also disable third-party browser extensions altogether from the advanced settings. Connecting through a VPN or running your browser within a virtual machine can help protect your identity and the browser.

To secure a web browser, consider using various browsers and implementing Group Policies from a server (domain controller) to adjust security settings for many computers across the network. Removing unnecessary applications is important, but removing web browsers can be difficult or impossible. Keep in mind that the technology world is changing quickly, so periodically review your security policies and keep up to date with the latest browser functionality, updates, security settings, and malicious attacks.

Securing Other Applications

Typical users should only have access to applications they need, such as Command Prompt and PowerShell in Windows. Protective measures should be put in place to prevent unauthorized access and user error. One way to do this is to use User Account Control (UAC) on qualifying Windows operating systems, which keeps every user in standard user mode instead of an administrator with full administrative rights. This is meant to prevent unauthorized access and avoid user error in the form of accidental changes.

Another way to deny access to applications is to create a policy, such as Don't Run Specified Windows Applications or Run Only Specified Windows Applications policy. This can be found at the Marketing-Policy in Windows Server. For applications that users are allowed to work with, they should be secured accordingly. This includes updating, patching, or having the appropriate service packs installed. Mobile apps should also be secured, with strong passwords for email accounts and accounts to "app stores" and similar shopping portals.

Back-office applications that run on servers should be hardened and secured, such as database software, email server software, and FTP servers. These applications have their own set of confidentiality requirements and might be insecure out-of-the-box. Renaming accounts and establishing strong passwords are essential steps to protect these systems.

Organizations use many applications specific to their type of business, so it is important to stay on top of vendors that supply updates and new versions of software. Always test the effects of one new piece of software on the rest of the installed software before implementation.

Secure Programming

A Must Know!

Many applications are inherently insecure, but implementing secure programming or coding concepts can limit this issue. These best practices harden the code of the application, preventing vulnerabilities and demonstrating how to defend against them.

Software Development Life Cycle

The software development life cycle (SDLC) is an organized process of planning, developing, testing, deploying, and maintaining systems and applications. The waterfall model is a common SDLC model used by companies, broken down into several sequential phases. The SDLC includes planning and analysis, software/systems design, implementation, testing, integration, deployment, and maintenance. The waterfall model is similar to other models, such as the V-shaped model, which emphasizes testing, and the agile model, which breaks work into small increments and is more adaptable to change.

DevOps, a term for software development and information technology operations, emphasizes collaboration between the two departments to make coding, testing, and releasing of software more efficient and secure. A secure DevOps environment should include secure provisioning and de-provisioning of software, services, and infrastructure, security automation, continuous integration, baselining, infrastructure as code, and immutable systems. Immutable systems mean that software and services are replaced instead of being changed, and rapid, efficient deployment of new applications is at the core of DevOps.

In summary, the SDLC is a crucial tool for developers to ensure secure applications throughout their project lifecycle. Different models, such as the waterfall, V-shaped, RAD, and agile, can be used depending on the organization and the number of systems involved.

Core SDLC and DevOps Principles

The core fundamentals of security are essential for programmers and security administrators to maintain confidentiality, preserve integrity, and protect availability. These concepts are crucial in a secure code review, which is often included as part of the testing phase of the Software Development Life Cycle (SDLC). Quality assurance policies and procedures should be implemented during the development and testing of code, including procedures, checks and balances, and extensive documentation.

Treatment modeling is a secure and structured approach that organizations use to prioritize threats to an application based on their potential impact. This process includes identifying assets, uncovering vulnerabilities, documenting threats, and rating them according to their potential impact. Threat modeling is often incorporated into the SDLC during the design, testing, and deployment phases.

Other important security principles include the principle of least privilege, defense in depth, user input validation, minimizing attack surface area, establishing secure defaults, providing authenticity and integrity, failing securely, and fixing security issues correctly. Other concepts to consider include o

For the Security+ exam, the most important SDLC phases are maintenance and testing. Maintenance involves updating applications, correcting failures, and constantly monitoring. Testing involves checking for bugs and errors, understanding vulnerabilities and attacks, and implementing patches to fix problems without causing other issues or application regression. By following these principles, developers can ensure the security of their systems and applications.

NOTE

Input validation is crucial for website design and application development, as it ensures the correct usage of data and prevents security vulnerabilities such as sensitive data exposure and data corruption. Data can be validated in various ways, including coded data checks, consistency checks, spelling and grammar checks, and more. Input validation should be done both on the client side and server side. For example, a PHP-based contact form should be checked for errors or malicious input, such as a backslash in the domain name. Server-side validation is even more important, as any data passed on by the form should be checked at the server. This can be done within the database software or through other means. Input validation is the key to preventing attacks such as SQL injection and XSS. All form fields should be tested for good input validation code, both on the client and server side, to develop complete mediation of requests. To prevent sensitive data exposure, inputted data should be stored or transmitted in clear text, use strong encryption, HTTPS for authentication, and use a salt to strengthen hashed passwords.

Programming Testing Methods

The text discusses various testing methods for identifying programming vulnerabilities and preventing attacks, including system testing, input validation, and fuzzing, which when combined during the SDLC phase increase the likelihood of a secure application.

White-box and Black-box Testing

System testing is divided into two categories: black-box and white-box. Black-box testing involves testing the system's functionality without specific knowledge of the system code or programming. It often aims to crash the program, indicating a lack of error-handling code or input validation. White-box testing, or transparent testing, involves testing the internal workings of the application or system, requiring programming knowledge and detailed information about the system's design. Techniques like fuzzing, data flow testing, stress testing, penetration testing, and sandboxes are used. Gray-box testing, a more recent category, involves testing at the user level, using internal knowledge of the system to create tests.

Compile-Time Errors versus Runtime Errors

Programmers and developers should test for potential compile-time and runtime errors. Compile-time errors, such as syntax and type-checking errors, can be checked without running the program. Runtime errors occur while the program is running and can include memory issues, invalid parameter values, or buffer overflows. These errors can lead to software exceptions. Structured exception handling (SEH) is a mechanism used to handle both types of exceptions, allowing programmers to control exception handling and support debugging. Security concerns are more focused on the runtime environment, as it is often exploited by attackers.

Static and Dynamic Code Analysis

Static code analysis is a debugging method that examines code without executing the program, using visual or automated tools. It can identify major issues and potential disasters. Dynamic analysis, like fuzz testing, is used to identify minor defects and vulnerabilities. This combination is known as glass-box or white-box testing.

Fuzz Testing

Fuzz testing is a method where random data is inputted into a computer program to identify vulnerabilities, often without knowing the program's source code. This process is commonly used to check file parsers in applications like Microsoft Word and network parsers used by protocols like DHCP. The test can uncover system failures, memory leaks, and error-handling issues. Once the test is complete, the code is reviewed, strengthened, and vulnerabilities removed. Reusing code is recommended to avoid common mistakes and ensure it is up-to-date and approved for use.

Programming Vulnerabilities and Attacks

This section discusses program code vulnerabilities and attacks, providing mitigation strategies during application development. It is not a programming seminar but a brief description of attack methods and defenses. CompTIA expects a broad understanding of these methods.

Backdoors

Backdoors are used in computer programs to bypass security mechanisms. To avoid these, update the operating system, applications, and firmware on devices, and carefully check program code. Malicious individuals can install code through backdoors, such as reactivating user accounts with a logic bomb. To prevent this, ensure the OS is updated and consider job rotation.

Memory/Buffer Vulnerabilities

Memory and buffer vulnerabilities are common issues in computer systems, with buffer overflow being the most significant. This occurs when a process stores data outside the intended memory, potentially leading to erratic behavior in the application. Stacks and heaps can be affected by buffer overflows, which can be initiated by certain inputs or corrupted by no-operation machine instructions. To prevent this, patch the system or application, ensure data execution prevention, and use bounds checking.

Integer overflows occur when arithmetic operations attempt to create a numeric value too big for available memory space, creating a wrap that can cause resets and undefinable behavior in programming languages like C and C++. Memory leaks are caused when a program does not release memory properly, reducing performance and potentially causing serious system stability issues. Memory debuggers can check for leaks, but garbage collection libraries should be added to C, C++, or other programming languages.

Pointer dereferencing, such as null pointer dereference, is another potential memory-related issue. Programs that contain null pointer dereference generate memory fault errors (memory leaks). To prevent this, meticulous coding and using special memory error analysis tools can help detect and correct the code causing the error(s). However, this concept can be used to attack systems over the network by initiating IP address to hostname resolutions, causing a return null.

Additionally, address space layout randomization (ASLR) can help prevent the exploitation of memory corruption vulnerabilities by randomly arranging different address spaces used by a program or process. While some exploits can bypass ASLR, many systems employ some version of ASLR.

Arbitrary Code Execution/Remote Code Execution

A Must Know!

Arbitrary code execution (RCE) is an attack where an attacker gains control of a target computer through a vulnerability, allowing them to execute commands at will. This can be done through browser URLs or the Net-cat service. To defend against RCE, applications should be updated, fuzz testing, and strong input validation are used. PHP on web servers can be set to disable remote execution, and web servers can block access from specific hosts. RCE is common with web browsers.

XSS and XSRF

Cross-site scripting (XSS) and cross-site request forgery (XSRF) are two web application vulnerabilities that can be exploited. XSS holes are vulnerabilities that can be exploited through code injection, where an attacker inserts malicious scripts into a web page to gain privileges and access to session cookies and other user information. To defeat XSS attacks, programmers can use output encoding, prevent HTML tags, and implement input validation. On the user side, increasing cookie security and disabling scripts can reduce the risk. XSRF attacks exploit the trust a user's browser has in a website, transmitting unauthorized commands. Reducing the chances of these attacks can be achieved by requiring tokens on web pages, using special authentication techniques, and submitting cookies twice.

More Code Injection Examples

Code injection is a common issue in databases, particularly relational databases like SQL and MySQL. These systems are typically written in the Structured Query Language (SQL), which can be used to execute improperly, gaining access to resources or changing data. SQL injection attacks can occur in databases, ASP.NET applications, and blogging software that use MySQL as a back end.

To defend against code injection, it is essential to constrain user input, filter user input, and use stored procedures such as input validating forms. NoSQL databases, which do not use SQL or use code in addition to SQL, are vulnerable to brute-force attacks and connection pollution. To protect against NoSQL injection, it is crucial to carefully scrutinize the service level agreement (SLA) between the company and the cloud provider.

LDAP injection is similar to SQL injection, using a web form input box to gain access or exploiting weak LDAP lookup conditions. The Lightweight Directory Access Protocol (LDAP) is used to maintain information directories, and strong input validation is the best way to protect against this.

XML injection attacks can compromise the logic of XML applications, potentially obtaining administrative access. To prevent this, it is essential to filter in allowed characters and use "default deny" to restrict input validation.

Code injection techniques involve adding or inserting code into existing code or executing system-level commands on vulnerable applications or OS. DLL injection is another type of injection attack that can influence program behavior. Input validation is the best way to defend against these techniques during the development, testing, and maintenance phases of the SDLC.

A Must Know!

Directory Traversal

Directory traversal, also known as the. "/ Attack," is a method to access unauthorized parent or root directories, often used on PHP servers on Linux or UNIX-based web servers. It can be perpetrated on Microsoft operating systems. Preventing this can be achieved by updating the OS or checking code for vulnerabilities.

Zero Day Attack

Zero day attacks are attacks that exploit software vulnerabilities before the creator can patch them. These attacks can cause damage even after the creator is aware of the vulnerability. To prevent zero day attacks, newer operating systems with protection mechanisms, multiple firewalls, and whitelisting are recommended. The CompTIA Security+ exam objectives require basic knowledge of programming languages and methodologies, including Visual Basic, C++, C#, Java, Python, HTML, ASP, PHP, and SQL. This knowledge will help in the exam and as a security administrator.

CHAPTER 6:

ACRONYMS	LIST OF ABBREVIATIONS
LAN	*Local Area Network*
WAN	*Wide Area Network*
ISO	*International Organization for Standardization*
ANSI	*American National Standards Institute*
OSI	*Open Systems Interconnection*
TCP/IP	*Transmission Control Protocol/Internet Protocol*
MAC	*Media Access Control*
CAM	*Content Addressable Memory*
DHCP	*Dynamic Host Configuration Protocol*
ARP	*Address Resolution Protocol*
STP	*Spanning Tree Protocol*
CSU	*Channel Service Unit*
DSU	*Data Service Unit*
DCE	*Data Communications Equipment*
DTE	*Data Terminal Equipment*
VLAN	*Virtual Local Area Network*
CIDR	*Classless Inter-Domain Routing*
VLSM	*Variable-Length Subnet Masking*
IEEE	*Institute of Electrical and Electronics Engineers*
NAC	*Network Access Control*
PAT	*Port Address Translation*
PBX	*Private Branch Exchange*
PSTN	*Public Switched Telephone Network*
VoIP	*Voice over Internet Protocol*
QoS	*Quality of Service*
SSH	*Secure Shell*
SIP	*Session Initiation Protocol*
SaaS	*Software as a Service*
ASP	*Application Service Provider*

IaaS	*Infrastructure as a Service*
PaaS	*Platform as a Service*
SECaaS	*Security as a Service*
MaaS	*Monitoring as a Service*
XaaS	*Anything as a Service*
PKI	*Public Key Infrastructure*
CASB	*Cloud Access Security Broker*
LDAP	*Lightweight Directory Access Protocol*
DoS	*Denial of Service*
DDoS	*Distributed Denial of Service*
XSS	*Cross-Site Scripting*
SMTP	*Simple Mail Transfer Protocol*
DLP	*Data Loss Prevention*
SSL	*Secure Sockets Layer*
TLS	*Transport Layer Security*
VPN	*Virtual Private Network*
CVE	*Common Vulnerabilities and Exposures*
FTP	*File Transfer Protocol*
FTPS	*FTP Secure*

Chapter 6

Network Design Elements

Key Topic!

Network design is crucial for a secure computing environment, defending against various types of network attacks. Identifying and addressing network threats is the next step in securing your network. The security administrator's responsibility may vary with cloud inclusion, depending on third-party provision and private domain ownership. They must understand computer networking functions and have working knowledge of networks and CompTIA Network+ certification. This book will focus on the security side of things as it progresses.

Network Design

A Must Remember!

Network design is crucial for security, protecting network devices and clients from attacks. Implementing address translation, using standard private IP ranges, and understanding network zones are essential. Utilizing subnetworks, VLANs, network access control, and secure telephony devices can enhance network design. The OSI model, a prerequisite for the Security+ exam, can help stimulate brain thinking and enhance network security.

The OSI Model

The Open Systems Interconnection (OSI) reference model, created by ISO and represented by ANSI in the US, aims to facilitate interconnection between systems.

1. Explain network connections between LAN and WAN hosts.
2. Present a technique for categorizing communication protocol suites (such as TCP/IP).
3. Demonstrate how various protocol suites may communicate with one another.

The OSI model categorizes existing network communications into seven layers, allowing engineers to understand communication protocols. It is an abstract way of categorizing existing network communications.

Layer #	NAME	USAGE	UNIT OF MEASURE
Layer 1	Physical layer	Physical and electrical medium for data transfer	Bits
Layer 2	Data link layer	Establishes, maintains, and decides how data transfer is accomplished over the physical layer.	Frames
Layer 3	Network layer	Dedicated to routing and switching information between different hosts, networks, and internetworks	Packets
Layer 4	Transport layer	Manages and ensures error-free transmission of messages between hosts through logical addressing and port assignment (connection-oriented). Also manages streaming connections, where *n* number of errors are permitted (connectionless).	Segments (TCP) Datagrams (UDP)
Layer 5	Session layer	Governs the establishment, termination, and synchronization of sessions within the OS over the network and between hosts.	Messages

Note: The table's "units of measurement" are commonly known as protocol data units, or PDUs.

Understanding the OSI model is crucial for the Security+ exam, so understanding its layers, order, and basic descriptions is essential. Chapter 7 discusses applying protocols to OSI layers. Additional resources include computer networking books, online articles, and networking training classes.

Understanding the OSI model is crucial for the Security+ exam, so understanding its layers, order, and basic descriptions is essential. Chapter 7 discusses applying protocols to OSI layers. Additional resources include computer networking books, online articles, and networking training classes.

Network Devices

A Must Remember!

Central connecting devices like switches and routers are crucial for network security. Attacks on these devices could disrupt the entire LAN. To protect these devices from both internal and external attackers, it's essential to expand your thinking beyond IP addresses on the Internet.

Switch

Ethernet switching, developed in 1996, replaced hubs and bridges due to improved data transfer and security. Switches connect all computers on a network, mapping MAC addresses to physical ports, securing the network and increasing data throughput. The CompTIA Security+ exam focuses on layer 2 and layer 3 switches, with most attacks being layer 2-based. It's essential to secure both layers for optimal network performance.

Although the switch is a good star-based topological solution, it does have significant security considerations. Among these include, but are not limited to, the following:

MAC Flooding

Switches store MAC addresses in the Content Addressable Memory table (CAM table). A MAC flood can send multiple packets, reducing network bandwidth and allowing unauthorized access to data. Some switches can shut down a port if it receives a certain amount of packets with different source MAC addresses. Cisco uses port security to restrict access to ports by limiting and identifying MAC addresses. Other security measures include network access control, dynamic VLANs, network intrusion detection systems, and consistent network monitoring.

MAC Spoofing

MAC spoofing involves an attacker masking a computer's MAC address with another number, allowing a user to trick a switch and gain access. Cisco's port security features and NAC systems can help prevent this. DHCP starvation attacks can be enhanced by broadcasting spoofed MAC addresses, exhausting address space. To prevent this, enable DHCP snooping and close untrusted switch interfaces. ARP spoofing can be prevented by reducing entry time, checking static entries, dynamic ARP inspection, and DHCP snooping.

Note: Remember to save and back up the relevant configuration files whenever you make changes to any switch, including Cisco switches. Maintain a record of the modifications made to each backup file.

Physical Tampering

Switches with a dedicated management port can be vulnerable to attacks, including MAC flooding and looping. To prevent cable loops, consider a hierarchical switched environment with all LAN switches connected to a master switch. Enable loop protection within firmware, such as Spanning Tree Protocol (STP), to build a logical loop-free topology for the Ethernet network. Physically secure switches, check for internal cable loops, disable unused ports, and use good cable management to label ports and cables. Disable unused ports and organize patch cables as needed.

Bridge

A bridge separates a physical LAN or WLAN into two logical networks by physically connecting devices to both sections. It stores MAC addresses and decides whether to cross to the other side, eliminating broadcasting. Bridges are less common since switching, but they may still be used in wireless bridged environments. Security is crucial, and a newer, more secure solution is recommended.

Router

A router connects multiple networks to form an internetwork, functioning on the Network layer of the OSI model. They come in various forms, including SOHO routers, servers with multiple network adapters, and black-box devices like Cisco routers. Routers are intelligent and often have their own operating system, such as IOS.

Assaults such as denial-of-service assaults and malware incursions, which are discussed in greater detail later in this chapter, can target routers and propagate malware and attacks to other parts of the network. There are a few techniques to defend routers against these attacks:

- **Secure Router Configuration:** Most routers are inherently insecure, with default passwords, usernames, and IP addresses. To protect, create complex passwords and lock down all default configurations before putting the router on a live network.
- **Firewalls:** Firewalls, either individual devices or routers, protect against unwanted traffic. They can block specific types of traffic, such as ICMP, preventing pinging. More information on firewalls can be found in Chapter 8, "Network Perimeter Security."
- **Intrusion prevention systems (IPSs):** An IPS detects and prevents directed attacks, botnet attacks, and malware, and can be installed as a network-based solution or on specific computers and routers.
- **Secure VPN connectivity:** Virtual private networks (VPNs) provide secure connections using IPsec and SSL, and can be implemented using SOHO routers, VPN concentrators, advanced routers like Cisco, or Windows Server.
- **Content filtering:** Content filtering blocks or restricts access to specific websites, providing protection from malicious ones. It can be installed as a server, appliance, or router.
- **Access control lists (ACLs):** Access control lists (ACLs) are routers and firewalls that enable or deny traffic, blocking IP spoofing attacks. They can be implemented on physical devices and can be found in Chapter 11.

Note: The CompTIA Security+ objectives also refer to the channel service unit (CSU) and Data

Service Unit (DSU). These two devices—often combined as a CSU/DSU—are within the realm of data communications equipment (DCE). They connect data terminal equipment (DTE) such as a router to a digital circuit such as a T-1. Today, the functionality of these devices (or device) is often incorporated into a router in the form of a WAN interface card (WIC) or otherwise. Cable and DSL modems are also considered to be CSU/DSUs. It's important to update the firmware on these devices periodically and replace old hardware with new devices.

Quick Hits!

Network Address Translation, and Private versus Public IP

Network address translation (NAT) is a process that changes an IP address while it's in transit across a router, remapping a larger private address space to a single public one. Originally designed to address IPv4 address exhaustion, NAT now provides protection by hiding a person's private internal IPv4 address, known as the firewall effect. Advanced routers allow port address translation (PAT), which translates both IPv4 addresses and port numbers.

Private addresses are not displayed directly to the Internet and are typically behind a firewall or NAT-enabled device. They are assigned automatically to clients by SOHO routers or DHCP servers. Public addresses are displayed directly to the Internet and can be accessed by anyone worldwide. Most private IPv4 ranges are reserved for private use.

Types of IPv6 Addresses

IPv6 auto-configured addresses pose risks as they can be easily obtained without user interaction, potentially leading to insecure communication. To prevent this, use 802.1X authentication, encrypted tunnels, and certified network adapters for secure transmissions.

IP security is crucial for both IPv4 and IPv6 networks. IPv6 has built-in IPsec support, but it may not be the best method for your organization. IPv4 can use IPsec but lacks support, so alternative security methods may be needed. Design both protocols to address all security concerns and thoroughly test them on multiple platforms.

IPv6 Type	Address Range	Purpose
Unicast	Global unicast starts at 2000 Link-local ::1 and FE80::/10	Address assigned to one interface of one Host.
Any cast	Structured like unicast addresses	Address assigned to a group of interfaces on Multiple nodes. Packets are delivered to the "first" interface only.
Multicast	FF00::/8	Address assigned to a group of interfaces on Multiple Nodes. Packets are delivered to all Interfaces.

Network Zones and Interconnections

When designing your network, consider all its components and connections to other networks. Consider if you manage a single local area network or multiple networks, the number of internet connections, and the need for cloud or virtualization. Consider sharing information with remote employees and securing it from average users.

NOTE

LAN Versus WAN

A local area network (LAN) is a group of networked computers in a small space, usually assigned private IP addresses and behind a firewall. They usually connect to the Internet via a router using NAT. Ensuring security on LANs involves placing them behind the router, assigning private IP addresses, and installing anti-malware programs. A wide area network (WAN) covers a larger geographic area and requires telecommunications or data communications services. All connections should be monitored and filtered if possible.

Internet

The Internet, a global network of computers and networks, is vulnerable to attacks. To protect all types of sessions, use protected VoIP systems, virtual private networks, firewalls, and antimalware programs. For systems that access both the LAN and the Internet, create a demilitarized zone (DMZ) to separate the two.

Must Know!

Demilitarized Zone (DMZ)

A demilitarized zone (DMZ) is a network area that houses servers that host information on the Internet, including web, FTP, mail, and database computers. Each server must have a proper default gateway IP address for internet access. DMZs can be set up in various ways, such as the 3-leg perimeter DMZ, which connects WWW and FTP servers. A firewall can be secured on the DMZ connection (192.168.100.200) and the LAN connection (172.29.250.200). It should only have inbound ports 80 (WWW) and 21 (FTP) open, filtering inbound traffic. The LAN connection can be shielded on the inbound side.

DMZs can be created logically or physically, with back-to-back perimeter network configurations and router setups depending on the organization's network architecture and security concerns.

Intranets and Extranets

Intranets and extranets are used by companies to share data using the internet while maintaining security. Intranets involve only one company and can be simple or complex, while extranets involve multiple companies or users. Both types of networks share data and resources, but only specific types of information are stored. Security is a major concern, and proper authentication schemes are implemented to ensure only authorized users can access data. Confidential, secret, and top secret information should not be hosted on these networks. Deploying firewalls in advance is crucial for maintaining security. Separating data into separate physical locations can be more secure, but may be more expensive.

Network Access Control (NAC)

Network access control (NAC) is a secure method of governing network connections. It sets rules for antivirus protection and system updates, preventing malicious attacks. Client computers are denied access until they are updated, which can be automated by NAC solutions. Agents are either preinstalled software or scanned remotely. There are two types: persistent and dissolvable. Some companies offer hardware-based NAC solutions, while others offer paid software-based and free options. The IEEE 802.1X standard, or port-based network access control, is a basic form of NAC.

Sub-Netting

Sub-Netting is the process of creating subnetworks by manipulating IP addresses, which are distinct parts of a single IP network. Sub-Netting is used for several purposes:

- It separates the network into sections, which improves security.
- The utilization of IP address space is more efficient.
- It lessens crashes and broadcast traffic.

To avoid security vulnerabilities, network administrators should start with subnet 1, such as 192.168.50.16, instead of subnet 0 (zero). This prevents confusion and ensures hosts on the first subnet don't have access to more of the network than they should.

Classless inter-domain routing (CIDR) is a method used by organizations to subnet their network, making it classless by changing the subnet mask to a different subnet. This is based on variable-length subnet masking (VLSM). For example, a Class 'A' network with a subnet mask of 255.255.255.224 can be sub-netted as a Class C network, allowing 30 usable hosts per subnet. This ensures proper communication between different subnets.

To avoid security vulnerabilities, network administrators should start with subnet 1, such as 192.168.50.16, instead of subnet 0 (zero). This prevents confusion and ensures hosts on the first subnet don't have access to more of the network than they should.

To avoid security vulnerabilities, network administrators should start with subnet 1, such as 192.168.50.16, to avoid confusion about the actual network number and its subnets. Using a default subnet mask can create security vulnerabilities, as hosts on that subnet may access more of the network than they normally should. Classless inter-domain routing (CIDR) is another common example of sub-netting a network, making it classless by changing the subnet mask. For example, a 'Class A' network with a subnet mask of 255.255.255.224 can have 30 usable hosts per subnet.

Sub-netting allows organizations to assign departments to subnets, assigning varying security policies. Incidents and attacks are usually isolated to the subnet, so routers should be regularly updated and monitored.

Virtual Local Area Network (VLAN)

A VLAN is a network segmentation device that organizes, reduces collisions, and enhances performance. It can be controlled by a switch and can be set up physically, like a port-based VLAN. It can isolate traffic and be associated with specific computers, ensuring secure network jacks for VLANs with confidential data access.1

VLANs, including protocol-based and MAC address-based types, require separate security precautions beyond the CompTIA Security+ exam. IEEE 802.1Q is the most common standard, modifying Ethernet frames with appropriate VLAN information. VLAN hopping bypasses restricted access to network resources.

Types of VLAN Hopping

VLAN Hopping Method	How It Works	How to Defend
Switch spoofing	To mimic the VLAN trunking switch, the attacker computer has to be able to speak the trunking and tagging protocols. The attacker computer can then access traffic for one or more VLANs if the attack is successful.	Switch ports that are not in use should be placed in an unused VLAN. Statistically determine which switch ports are responsible for explicitly forwarding certain tags and for transmitting tagged frames as trunks. If required, turn off the Dynamic Trunking Protocol (DTP). Steer clear of default VLAN names like VLAN or VLAN1.
Double tagging	An attacker host affixes two VLAN tags to the frames it broadcasts in a double-tagging attack. The first switch the frame comes into contact with removes the first correct header, after which the frame is transmitted. The second switch the frame comes into contact with can then see the second, fake header.	Update your software or firmware. Select an underutilized VLAN to be the native (or default) VLAN for all trunks; do not use it for any other purpose. In case of using several 802.1Q switches, think about reworking the VLAN.

MAC flooding attacks can be persistent on VLANs, but VLAN hopping isn't possible due to packet flow constraints. VLANs can also be targeted by ARP, brute-force, and spanning-tree attacks.

Telephony

Telephony, a primary communication method, requires various equipment, including older modems and computer telephony integration (CTI). Security administrators must ensure special telephones and servers are protected against new attacks, as they are crucial for maintaining communication.

Modems

Modems are often used by network administrators in NOCs or server rooms to connect to servers and networking equipment via dialup lines, often without authentication. Telnet can also be used, but this is insecure due to war-dialing, where computers scan telephone numbers one at a time, potentially allowing hackers to access computer networks without an authentication scheme.

To protect modem connections, network admins should use callback features, strong passwords, and use dial-up modems sparingly in secure locations. Telnet is not secure and should be replaced with SSH or other secure methods. For users using modems on client computers, set them to not answer incoming calls and avoid remote control software. Consider upgrading to a faster, more secure Internet access solution.

PBX Equipment

A private branch exchange (PBX) is a crucial component of an organization's internal phone connections and public switched telephone network (PSTN). Its security is a concern, as it can be exploited for free long distance service or social engineering. To secure a PBX, it should be in a secure room, change passwords regularly, and only allow authorized maintenance. PBX devices can also act as computer-telephony integration servers or incorporate VoIP, known as IP-PBX.

VoIP

VoIP is a communication protocol used for transmitting voice data over IP networks. It can be vulnerable to attacks like individual computers and servers, such as denial-of-service attacks. To protect VoIP servers, security administrators should implement similar precautions as traditional servers. Some solutions use the Session Initiation Protocol (SIP), which can be exploited by MIMO attacks. Regular updates, encryption, and authentication are essential. Quality of Service (QoS) can be used to prioritize traffic on routers, increasing IP telephony availability.

Cloud Security and Server Defense

The term "cloud" historically referred to the Internet and telephone company infrastructure, including central and switching points. Today, it is more service-oriented and widely used in marketing, with its meaning shifting from technical to more user-friendly. The term remains relevant in the technical sense, but its meaning has evolved over time.

Chapter 2 discusses basic computer protection, OS hardening and virtualization, and secure programming. The chapter then moves on to server defense, which is crucial in networking. Servers control data transmission and reception, including FTP, websites, email, and database formats. Many servers are now in the cloud, which is essential for networking.

Cloud Computing

Key Topic!

Cloud computing offers on-demand services that extend computer or organization's network capabilities. These can be free or pay-per-use, offering data access, storage, infrastructure, and online gaming. Network connections are required for real-time access.

Benefits include lowered costs, less administration, more reliability, increased scalability, and performance. For example, browser-based email can be used by small businesses without the costs of hosting a domain. However, there are administrative control and security concerns.

Services related to cloud computing may often be divided into many categories:

- **Software as a service (SaaS)**
  SaaS refers to third-party applications provided over the internet, often run through a web browser or remote desktop programs, such as webmail, without the need for local installation.
  Note: SaaS and the application service provider (ASP) paradigm are frequently contrasted. SaaS usually provides a broad service to a large number of consumers. On the other hand, an ASP usually serves a limited number of users with a service (perhaps a single application).

- **Infrastructure as a service (IaaS):** Organizations increasingly benefit from cloud-based services that offer computer networking, storage, load balancing, routing, and VM hosting.

- **Platform as a service (PaaS):** PaaS is a service that offers software solutions for organizations, enabling them to develop applications in a virtual environment without the cost or administration of a physical platform, often utilizing IaaS for infrastructure.

- **Security as a service (SECaaS):** Cloud Security Alliance (CSA) defines SECaaS as a service where a large provider integrates security services into a company's infrastructure, offering efficient and cost-effective solutions for businesses with limited IT staff or budget.1

  Note: New services will sometimes appear, such monitoring as a service (MaaS), a platform that makes it easier to run continuous monitoring in the cloud. Services that are cloud-based come in a variety of forms. They frequently fit into the "anything as a service" (XaaS) category if they don't belong in the preceding list.

Between 2005 and 2010, cloud services were slow to adopt due to security issues. However, after 2010, implementation increased dramatically. Managed service providers (MSPs) offer network, application, system, and management services using a pay-as-you-go model. Organizations employ three main kinds of clouds: communal, private hybrid, and public.

Let's talk about each briefly.

- **Public cloud:** Public cloud space, offered by service providers like Google, Rackspace, and Amazon, provides low-cost applications and storage over the internet, offering scalability and flexibility.

- **Private cloud:** The security administrator controls data and infrastructure, with limited access to the cloud, typically behind firewalls. Resources may come from third parties or the administrator's server room.

- **Hybrid cloud:** Hybrid scenarios combine public and private clouds, with dedicated servers within the organization and third-party servers, ensuring in-house storage of confidential data.

- **Community cloud:** Community clouds are a mix of public and private, allowing multiple organizations to share the public portion of their data storage and computing methods.

Organizations choose cloud types based on budget, security, and manpower. Private cloud options are often beyond their capabilities, leading to public or community-based options. However, resources must be secured, and the organization plays a role in this process.

Cloud Security

Cloud security depends on the level of control a security administrator has and the types of controls they implement. Key concerns include loss of physical control of data, privacy, accountability, improper authentication, data sensitivity, integrity, segregation, location, and recovery issues. Cloud service providers can also be abused, as attackers use their infrastructure for powerful attacks.

The following are some remedies for these security problems:

Complex passwords

As I shall emphasize several times in this article, strong passwords are crucial and far more vital than that. The following are examples of approved password schemes as at the time this book was written:

- **For Overall Security:** a minimum of 10 characters, with a capital letter, a number, and a special character at least

- **For Confidential Data:** a minimum of 15 characters, including at least two capital, two numeral, and two special characters.

A security administrator should use the second option for all cloud types, as public and private clouds are more secure, and a strong server-based policy is recommended for password enforcement.

- **Powerful Authentication Methods**

Multifactor authentication offers depth of defense, allowing for backups in case of compromise. It includes biometric confirmation, additional PIN, or smart card swipe. Physical feasibility depends on the cloud environment, but if possible, it should be considered.

- **Strong Cloud Data Access Policies**

To protect public clouds, define user access, resources, and permissions, use strong passwords and two-factor authentication, establish policies from servers, audit connected devices and apps, store different data types with different services, and approach data backing up as a separate procedure.

- **Encryption**

  Encryption of data files, whole disk encryption, and digitally signed virtual machines requires a robust public key infrastructure (PKI) for web browser access.

- **Standardization of Programming**

  Standardization in cloud application planning, design, programming, and execution is crucial, with input validation, fuzzing, and white-, gray-, or black-box testing being essential for consistency across platforms and programmers.

- **Protection of All the Data**

Data storage in multiple locations can be vulnerable, so it's crucial to maintain detailed documentation and implement cloud-based security controls. These controls include deterrent, preventive, corrective, and detective controls. These measures prevent data tampering, increase security, reduce tampering effects, and detect attacks in real-time, ensuring a robust defense plan.

The company is focusing on protecting user identity, privacy, and super-confidential information, including credit card numbers. They aim to secure physical servers, applications, and third-party cloud providers within contractual obligations. They also aim to promote data availability, ensuring that the design plan allows for secure access to available data even after security controls have been implemented.

Customers should demand transparency in cloud computing services and ensure compliance with their organization's security policies. When planning on on-premises, hosted, cloud, or a mix of these, consider using a cloud access security broker (CASB) to extend the reach of security policies beyond internal infrastructure.

Other "Cloud" Based Concerns

Cloud technologies, such as social media, file sharing, and P2P networks, can compromise security by allowing users to share personal information and data. Security administrators should consider whitelisting applications to protect users. P2P networks, which allow file sharing, gaming, and media streaming, can also introduce malware and potential system intrusion. Access to these networks requires a permanent "padlock" and a permanent "padlock" for users. Darknets, also known as F2F, create connections between trusted peers using nonstandard ports and protocols, making them difficult to detect. Employees or cloud providers may have trusted cloud-based resources to join a darknet, potentially leading to legal consequences. Thorough checks of cloud-based resources, employee screening, inspection of service-level agreements, and third-party IT auditors can help prevent these risks.

Server Defense

Servers are crucial for data storage, transfer, and access, requiring fast network connections and regular monitoring. They require regular audits and policy control. Servers exist in large and small networks, public and private clouds, and are often virtual. Five types of servers are of great importance, with threats and vulnerabilities and ways to protect them.

File Servers

File server computers store, transfer, migrate, synchronize, and archive files. Examples include Microsoft Windows Server, macOS/OS X Server, Linux server versions, and Unix. They are vulnerable to attacks and malware. To secure them, use hardening, updating, anti-malware applications, firewalls, HIDs, encryption, and regular monitoring.

Network Controllers

A network controller, like a Windows Server system, is a central repository for user accounts and computer accounts. It can be targeted by LDAP injection and Kerberos vulnerabilities, leading to privilege escalation or spoofing. To protect a Windows domain controller, it's essential to install specific security update hot patches, even if the latest service pack is installed.

E-mail Servers

E-mail servers, such as Microsoft Exchange, are vulnerable to various attacks, including XSS, overflows, DoS/DDoS, SMTP memory exploits, directory traversal attacks, and spam. Administrators must stay updated on the latest vulnerabilities and be prepared to shut down or quarantine the server at any moment. To protect email-based data, administrators should consider DLP and encryption, secure S and SMTP, encrypted SSL/TLS connections, security posture assessment (SPA), secure VPNs, and other encryption types, especially for web-based connections. Web-based connections can be particularly insecure, and administrators should consider moving away from Microsoft and towards Linux solutions like Java-based SMTP servers built into Apache or third-party tools like Zimbra. While Linux has not been attacked as often as Microsoft, the difference in attacks experienced has significantly decreased since the turn of the millennium.

Web Servers

Web servers, such as Microsoft's IIS, Apache HTTP Server, light tpd, Oracle iPlanet Web Server, and Sun Java System Web Server, are frequently targeted by various attacks. These include DoS attacks, overflows, XSS, XSRF, remote code execution, and backdoor attacks. Security administrators should regularly update their knowledge of Common Vulnerabilities and Exposures (CVE) and implement hardware-based firewalls, HTTPS, and automated vulnerability scanning programs. Apache web servers can also be targeted by attacks such as privilege escalation, code injection, and exploits. Patches to known CVEs should be applied promptly.

Security admins must be vigilant for Darkleech, a malicious Apache module that can initiate attacks, deliver malware and ransomware, or redirect users to other sites with exploit kits, although it's not limited to Apache, with most infected sites being Apache-based.

To combat Darkleech, webmasters can check for PHP files in suspiciously long hexadecimal names, filter out iframes, update the Apache server, and take it offline while it's repaired. In some cases, rebuilding the server or virtual server image is necessary.

FTP Server

An FTP server is a tool used to provide basic file access publicly or privately. It can be built into IIS, Apache FtpServer, and third-party offerings like FileZilla Server and Pure-FTPd. However, the default FTP server is insecure due to its use of well-known ports (20 and 21), lack of encryption by default, and basic username/password authentication. This makes it vulnerable to various types of attacks, such as bounce attacks, buffer overflow attempts, and attacks on the anonymous account.

To protect data stored on the FTP server, security administrators should consider incorporating FTP software that uses secure file transfer protocols like FTPS or SFTP, or using FTP software that dynamically assigns data ports instead of using port 20 every time. Encryption can prevent most attackers from reading data files, even if they can access them. If not a public FTP, the anonymous account should be disabled.

Another threat is the web shell, a program installed on a web server by an attacker, which can remotely access and reconfigure the server without the owner's consent. Web shells are remote access Trojans (RATs) and backdoors, offering an alternative way for attackers to access the website. To prevent this, security administrators should increase password security, eliminate unnecessary accounts, separate the FTP server and web server to different computers or virtual machines, and set up automated scans for web shell scripts. If a web shell attack is successful, security administrators must either search for and delete the original RAT files or re-image the system and restore from backup.

This summary covers various servers, including DNS, application, virtualization, firewall/proxy, database, print, remote connectivity, and computer telephony integration. If you're responsible for securing a server, review CVEs and bulletins and be prepared to hot-patch the system. This involves having an RDP, VNC, or other remote connection ready on your desktop for quick access to the server.

CHAPTER 7:

ACRONYMS	LIST OF ABREVIATIONS
FTP	*File Transfer Protocol*
TCP	*Transmission Control Protocol*
UDP	*User Datagram Protocol*
HTTP	*Hypertext Transfer Protocol*
SMTP	*Simple Mail Transfer Protocol*
SSH	*Secure Shell*
DHCP	*Dynamic Host Configuration Protocol*
POP3	*Post Office Protocol 3*
TLS	*Transport Layer Security*
ICMP	*Internet Control Message Protocol*
DoS	*Denial of Service*
TSIG	*Transaction Signature*
ARP	*Address Resolution Protocol*
VLAN	*Virtual Local Area Network*
DDoS	*Distributed Denial of Service*
IDS	*Intrusion Detection System*
IPS	*Intrusion Prevention System*
DNS	*Domain Name System*
FTPS	*FTP Secure*
SFTP	*Secure File Transfer Protocol*
SCP	*Secure Copy*
URL	*Uniform Resource Locator*
MAC	*Media Access Control*
WWN	*World Wide Name*
MITM	*Man-in-the-Middle*
MITB	*Man-in-the-Browser*
SSL	*Secure Sockets Layer*
CHAP	*Challenge-Handshake Authentication Protocol*
NetBIOS	*Network Basic Input/Output System*
LAN	*Local Area Network*
DNSSEC	*Domain Name System Security Extensions*

Chapter 7

Networking Protocols and Threats

Servers need open ports for client connections, but every open port is a vulnerability. Understanding common protocols and protecting against threats like denial-of-service, man-in-the-middle, replay, and session hijacking is crucial. Building knowledge of ports and protocols can help protect servers and networks from various attacks. It's important to stay updated with security bulletins, Common Vulnerabilities and Exposures (CVEs), and security updates to stay updated.

Ports and Protocols

Key Topic!

Secure a host's ports and protocols as they are the gateways into an operating system, and security administrators must monitor, audit, and implement updated defense mechanisms to combat malicious attacks, understanding ports and protocols is crucial.

Port Ranges, Inbound Versus Outbound, and Common Ports

Ports are crucial in securing hosts and are often used on exams. They act as logical communication endpoints for computers, with protocols such as TCP and UDP controlling these ports. There are 65,536 ports, divided into categories: Inbound ports, which are used when another computer wants to connect to a service or application running on your computer, and outbound ports, which are used when your computer wants to connect to a service or application running on another computer.

A security administrator should be most concerned with the inbound ports, as they relate to the applications, services, and protocols that run on a computer. The bulk of these protocols are application layer protocols, such as HTTP, FTP, SMTP, SSH, DHCP, and POP3. Application service ports are associated with these ports, and some protocols only use TCP transport layer connections or UDP only.

The IP address of a computer and the port number it is sending or receiving on are combined together to form a network socket address. For example, the initial FTP connection occurs when a user connects to an FTP server with FTP client software, while subsequent data connections are normally done on the server side via dynamically assigned ports. To secure FTP communications, consider using software that enables dynamically assigned ports during data transfers, such as Pure-FTPd on the server side and FileZilla on the client side.

For HTTP connections, consider using port 80 on a different server and port 110 on POP3 connections. Always be mindful of securing connections, as encrypted POP3 connections using SSL/TLS would likely use port 995.

Ports play a crucial role in network security, especially on routers and firewall devices. These devices operate on an implicit deny concept, denying all traffic unless a rule is made to open the port associated with the desired traffic type. To detect open ports, scan servers, routers, and firewall devices using tools like net stat, Nmap, or an online scanner. Unnecessary ports should be closed by recanting the host, making necessary changes in documentation, and following change management procedures.

Port zero security is another important aspect of network security. Only 65,535 of the 65,536 ports can be exploited due to their redirection to another dynamically assigned port. This is because programmers use port zero as a wildcard port, designing their applications to ask the operating system to assign a non-zero port. This means that only 65,535 of the 65,536 ports can be exploited.

Port zero may become more of a security concern with the growth of legitimate raw socket programming, which directly programs to network ports, bypassing the transport layer. Historically, raw sockets have been used by attackers to perform TCP reset attacks, setting the reset flag in a TCP header to 1, telling the computer to kill the TCP session immediately.

Protocols That Can Cause Anxiety on the Exam

Many protocols, including FTP, can be confusing due to their similar appearance and behavior. To make FTP sessions more secure, users can use FTP software that randomizes ports for transfers, or choose passive mode instead of active mode. Secure FTP (SFTP) and FTP Secure (FTPS) are examples of FTP protocols that are secured through encryption. SFTP uses SSH port 22 for connections, while FTPS uses SSL or TLS and uses ports 990 and 989 for secure connections and data transfers. FTPS can work in explicit or implicit modes, with explicit mode being more secure. Secure Copy (SCP) is another file transfer program that uses an additional protocol and its corresponding port for security. Understanding these acronyms can help alleviate confusion and help prepare for exams.

Malicious Attacks

A Must Know!

This section focuses on defining various types of malicious network attacks, highlighting their similarities and challenges in distinguishing them. It provides definitions, examples, and mitigating techniques for each attack.

DoS

Denial-of-service (DoS) is a term used to describe various network attacks that attempt to make computer resources unavailable. These attacks can be implemented in various ways, such as flood attacks, ping floods, Smurf attacks, and Fraggle attacks. Flood attacks involve sending many packets to a single server or host to disable it, while ping floods attempt to use up all available bandwidth. To deter this attack, ensure the system does not respond to ICMP echoes. Smurf attacks broadcast ICMP echo requests to every computer on its network or subnetwork, targeting a spoofed IP address. To protect against this attack, ensure hosts do not respond to pings or ICMP echoes, routers do not forward packets directed to broadcast addresses, sub netting with smaller subnetworks, and employing network ingress filtering. Fraggle attacks send traffic directed to port 7 (Echo) and port 19 (CHARGEN), and can be automated and modified using the Fraggle.c exploit code. SYN flood attacks involve sending a large amount of SYN request packets to a server to deny service. To defend against this, implement flood guards, recycle half-open connections after a predetermined amount of time, and use intrusion detection systems (IDSs) to detect the attack.

The Christmas attack, also known as the Christmas Tree attack or TCP Xmas Scan attack, can cause routers and devices to deny service or reboot. It is based on the Christmas Tree packet, which can be generated by various programs. If sent multiple times, it could potentially result in a DoS attack. Most routers and devices today block this type of packet, but an IDS/IPS solution can detect and prevent it from denying service.

Ping of Death (POD) is an older attack that sends an oversized and malformed packet to another computer. It is older and most firewalls block it before entering a network. To protect against POD, ensure hosts don't respond to pings or ICMP echoes, run the latest service packs and updates, update firmware on hardware-based firewalls, and update software-based firewalls. POD and other DoS attacks are considered deprecated by the security community.

DDoS

A Must Know!

DDoS attacks are a type of cyber threat where a group of compromised systems attacks a single target, causing a DoS to occur at that host. These attacks often use a botnet, a large group of computers known as robots or bots, known as zombies. The attacker exploits a single vulnerability in a computer system and makes it the DDoS master, communicating with other systems in the botnet. They often load malicious software on many computers, launching a flood of attacks by all zombies with a single command. DDoS attacks and botnets are often associated with exploit kits and ransomware. To protect against DDoS attacks, organizations can implement anti-DDoS technology and emergency response from an outside source or their cloud-based provider. If a

DDoS attack is detected, they should call their ISP and request that the traffic be redirected.

Sinkholes and Blackholes

DDoS attacks can be effectively combated using various security measures, including sinkholes, blackholes, and blackhole lists. A sinkhole redirects traffic to no routable addresses, while a blackhole identifies spammers and malware-infested domains. A DNS blackhole list (DNSBL) is a published list of IP addresses used in spamming and DDoS attacks, which can be used to block malicious traffic on an organization's DNS server.

NOTE

Spoofing attacks involve an attacker falsifying information to disguise themselves as another person. There are various types of spoofing, including man-in-the-middle attacks, internet protocols, URL spoofing, and phishing. Defending against these types is difficult, but by selecting and updating applications, and through user awareness, spoofing can be minimized. Uniform Resource Locators (URLs) and IP addresses can also be spoofed. IP address spoofing involves creating IP packets with a forged source IP address in the header, concealing their origin. Packet filtering and sessions that repeat authentication can defend against this type of spoofing. MAC addresses can also be spoofed, as they are more difficult to modify than IP addresses. World Wide Names (WWNs) are unique identifiers for SAS, ATA, and Fibre Channel equipment common to storage area networks (SANs). Proper authentication is beneficial in combating these types of spoof attacks. By carefully selecting and updating applications and using newer operating systems and network devices, organizations can mitigate risks associated with spoofing.

NOTE

Session hijacking is the exploitation of a computer session to gain unauthorized access to data, services, or other resources. It can occur at various levels, including application layer, network layer, and network layer. Session theft can be achieved through packet header manipulation or stealing a cookie from the client computer, which authenticates the client computer to a server. This can be achieved using encryption and long random numbers for the session key, or by requiring clients to periodically re-authenticate. TCP/IP hijacking is a common type of session hijacking, where an attacker takes over a TCP session between two computers without the need for a cookie or host access.

Blind hijacking occurs when an attacker blindly injects data into a data stream without being able to see whether the injection was successful. Clickjacking occurs when a user is tricked into clicking something different than what they thought they were clicking, usually implemented as a concealed link. Man-in-the-middle (MITM) attacks intercept all data between a client and a server, using cryptographic protocols such as Secure Sockets Layer (SSL) and Transport Layer Security (TLS).

Man-in-the-browser (MITB) attacks use a Trojan to infect a vulnerable web browser and modify web pages and online transactions, attempting to steal money or data. This can be prevented by updating the web browser, using transaction verification, and updating anti-malware on the computer in question.

Watering hole attacks are targeted attacks where an attacker exposes websites that the intended victim accesses and scans them for vulnerabilities. If the attacker locates a compromised site, the website is injected with JavaScript or other code injections designed to redirect the user when the user returns to that site. To prevent these attacks, organizations

Replay

A replay attack is a network attack where a valid data transmission is maliciously or fraudulently repeated or delayed. It differs from session hijacking, as the original session is intercepted and analyzed for later use. An attacker can impersonate the entity that sent the data, such as a bank using SSL or TLS to secure login sessions. Organizations can defend against replay attacks by using session tokens, timestamping, and timestamped nonces. CHAP-based authentication protocols can also provide protection against replay attacks.

Null Sessions

Null session attacks exploit NetBIOS ports 139 and 445 to make unauthenticated connections to a target computer. Successful attackers can find user IDs, share names, settings, and potentially gain access to files, folders, and resources. To protect against this attack, computers should be updated promptly and traffic on these ports should be blocked with a firewall or host-based intrusion prevention system. Command-line scripting can be used for both legitimate and illegitimate purposes, so operating systems should be updated and patched frequently.

Transitive Access and Client-Side Attacks

Transitive access is a method of attacking a computer based on the transitive property in mathematics. It involves a computer on the LAN trusting another computer, which can be manipulated. For example, if an attacker can create a trusted environment between two computers, they can compromise the database at computer C. This can lead to a client-side attack. Trusting relationships are often created to save time and bypass authentication methods. To prevent transitive access and client-side attacks, it is recommended to have every client computer authenticated when a session is started to another computer. This practice, along with firewalls, intrusion detection systems, and updates, is the best way to prevent such attacks.

DNS Poisoning and Other DNS Attacks

Quick Hits!

DNS poisoning is the modification of name resolution information in a DNS server's cache, redirecting client computers to incorrect websites. This can occur through improper software design, misconfiguration of name servers, and maliciously designed scenarios exploiting the DNS system's open architecture. Countermeasures include using Transport Layer Security (TLS) and digital signatures, Secure DNS (DNSSEC), and patching the DNS server. Transaction Signature (TSIG) can provide authentication for DNS database updates, but the hashing procedure may not be secure enough for an organization.

Unauthorized zone transfers replicate the database that contains DNS data and can be stolen by attackers accessing various hosts by IP address. To defend against this, zone transfers should be restricted and audited, and vigilant logging and regular checking of DNS records can help detect unauthorized zone transfers.

Windows computers' hosts file can also be attacked, as it is used to translate or resolve hostnames to IP addresses. Attackers may attempt to hijack the host file to alter or poison it, or try to bypass DNS altogether. The best defense is to modify the computer's hosts file permissions to read-only.

Pharming attacks redirect traffic to bogus and potentially malicious websites. To prevent pharming, end users should turn on phishing and pharming filters within the browser, be careful of which websites they access, and check their local hosts files.

Domain name kiting is the process of deleting a domain name during the five-day grace period and reregistering it for another five-day period.

ARP Poisoning

The Address Resolution Protocol (ARP) is a protocol that resolves IP addresses to MAC addresses, which can be poisoned or spoofed. ARP poisoning is an attack that exploits Ethernet networks, allowing attackers to sniff frames of information, modify it, or stop it from reaching its intended destination. Defenses for ARP poisoning include VLAN segregation/VLAN separation, DHCP snooping, and the open-source program ArpON.

CHAPTER 8:

ACRONYMS	LIST OF ABBREVIATIONS
NAT	*Network Address Translation*
SPI	*Stateful Packet Inspection*
ACLs	*Access Control Lists*
HTTP	*Hypertext Transfer Protocol*
SMTP	*Simple Mail Transfer Protocol*
DHCP	*Dynamic Host Configuration Protocol*
DNS	*Domain Name System*
FTP	*File Transfer Protocol*
TLS	*Transport Layer Security*
SSL	*Secure Sockets Layer*
TCP	*Transmission Control Protocol*
UDP	*User Datagram Protocol*
IDS	*Intrusion Detection System*
IPS	*Intrusion Prevention System*
HIDS	*Host-based Intrusion Detection System*
NIDS	*Network Intrusion Detection System*
NIPS	*Network Intrusion Prevention System*
DM	*Demilitarized Zone*
XSS	*Cross-Site Scripting*
BYOD	*Bring Your Own Device*
IP	*Internet Protocol*
URL	*Uniform Resource Locator*
DHCP	*Dynamic Host Configuration Protocol*
WAF	*Web Application Firewall*
DLP	*Data Loss Prevention*
NIS	*NIS - Network Information System*
NGFW	*NGFW - Next-Generation Firewall*
GUI	*GUI - Graphical User Interface*
UTM	*UTM - Unified Threat Management*
NGFW	*NGFW - Next-Generation Firewall*
WIDS	*WIDS - Wireless Intrusion Detection System*
WIPS	*WIPS - Wireless Intrusion Prevention System*

Chapter 8

Firewalls, Intrusion Detection & Prevention Systems

Key topic!

The network border, often referred to as the network perimeter, is the subject of this chapter. Because it has the entrances that many attackers try to utilize, this should be the main emphasis of a network security administrator when it comes to network security.

The author compares building an impenetrable castle of knowledge to a network. They use the castle analogy to illustrate how a network can be compared to a castle, with tall walls, clear land, or a moat, and drawbridges. The walls are meant to keep the average person out, while the moat makes it difficult for people to cross. This represents a network intrusion prevention system that detects and eliminates threats.

The text compares building an impenetrable castle of knowledge to a network, comparing it to a castle with tall walls, clear land, or a moat. The walls are designed to keep the average person out, like a firewall in a computer network. The moat makes it difficult for people to sneak in, while drawbridges are network ports open to the network. Network security administrators can close these ports at any time. Traps for people, known as honeypots, can also be set in a network environment. To ensure legitimate traffic enters and exits the network, proxy servers can act as go-betweens for computers inside the network and the Internet servers.

The network perimeter is challenging to visualize in a network environment, as it is often located in a single server room or data center, or a hybrid of in-house and cloud-based locations. To better visualize your network, map it out on paper or create documentation using tools like Microsoft Visio.

Firewalls and Network Security

Quick Hits!

Firewalls are widely used in businesses and households to protect networks. They come in various types, including hardware and software firewalls. Free software-based and firmware-based firewalls are available for download. Other tools include proxy servers, honeypots, and data loss prevention devices. The Firewall is the most crucial element in a network.

Firewalls

A Must Know!

Chapter 3 discusses personal firewalls, while network-based firewalls are crucial for protecting networks. They are installed on server computers, standalone appliances, or work as a single function on a device. They are often seen as a "brick wall" between LANs and the Internet.

A firewall in a computer network is designed to prevent malicious attacks and slow down the spread of viruses. It often has NAT in operation and has a local address and an Internet address. By default, the IP address is shielded, preventing incoming traffic. A good security administrator should check the firewall's firmware and scan it with third-party applications or a web-based port scanning utility. Open or unshielded ports should be addressed immediately and the firewall rescanned for vulnerabilities.

Firewalls should be used as intended, excluding extraneous services like web or SMTP servers. They inspect traffic and allow or deny it based on administrator-set rules stored within access control lists (ACLs). These rules control inbound and outbound traffic. Input deny is a default rule found in ACLs, often added automatically to the end of ACLs. This concept helps reduce vulnerability by allowing or denying specific traffic types based on their presence in the firewall's rule set.

Firewall rules should be specific, specifying the transport layer protocol, exact port, and IP address for any computer's IP address. They should be cautiously written to prevent filtering required traffic, as demonstrated by the example of a firewall rule denying TCP any port 53.

SPI and NAT are two types of firewall technologies that are integrated into many of today's firewalls.

Nonetheless, you have to be cognizant of a few other categories of firewall methodologies:

- **Packet filtering:** The firewall inspects each packet passing through it, accepting or rejecting it based on rules. There are two types: stateless packet inspection (SPI) and stateful packet inspection (SPI). Stateless packet inspection is vulnerable to IP spoofing attacks, while stateful packet inspection monitors network connections and distinguishes legitimate and illegitimate packets.
- **NAT filtering:** NAT endpoint filtering filters traffic based on ports (TCP or UDP) through basic endpoint connections, matching incoming traffic to outbound IP address connections, or matching incoming traffic to corresponding IP address and port.
- **Application-level gateway (ALG):** The ALG applies security mechanisms to specific applications like FTP or BitTorrent, supporting address and port

translation and checking application traffic types. It discards Telnet packets, adding a powerful layer of security but potentially causing performance degradation.

- **Circuit-level gateway (CGL):** Circuit-level gateways, located at the session layer of the OSI model, apply security mechanisms during TCP or UDP connections, allowing packets to flow without further checking.

A firewall can be configured in various physical configurations, such as a back-to-back configuration or a 3-leg perimeter configuration, as discussed in Chapter 6. A firewall typically consists of one firewall, allowing all devices and computers to enter the network. Devices "behind" the firewall are considered "after" the firewall, while those "in front" are "before". This analogy helps understand the concept of "in front of" and "behind" in relation to data entering the network and devices. Firewall logs are crucial for detecting intrusions and should be checked first. Accessing and reading logs is essential. For instance, a blocked Gnutella packet can be identified by its association with an older P2P file-sharing network, Gnutella. These external computers are random clients trying to connect to anyone possible, highlighting the importance of firewall logs in security.

To block an IP address from connecting to your network, you can add it to an inbound filter or an ACL. Both host-based and network-based firewalls can fall into the category of application firewalls. These firewalls operate on the application layer of the OSI model, allowing for more in-depth monitoring of the network. Some network appliances, like Barracuda, Citrix, Fortinet, and Smoothwall, also have web application firewalls (WAFs) designed to protect HTTP sessions from XSS attacks and SQL injection. These firewalls control traffic associated with specific applications, unlike stateful network firewalls.

Note: A multi-homed connection is required for a firewall appliance to connect to many networks; this is accomplished by using multiple network adapters. If you choose to use a DMZ or similar perimeter security method, it may be dual-homed (two adapters) or it may have more, maybe three network adapters.

Firewalls are often used interchangeably with other devices, such as SOHO users' all-in-one network devices. To protect users, it's important to teach them about disabling SSID broadcasting and enabling MAC filtering. Disabling SSID broadcasting prevents wireless connection, making it vulnerable to attackers. Enabling MAC filtering restricts access to computers without listed MAC addresses, a powerful tool for securing network media and devices.

A firewall can also function as, or in conjunction with, a proxy server, which we address in the section that follows, adding to the confusion.

Proxy Servers

Very Important!

A proxy server serves as a go-between for clients, which are often found on a local area network, and the servers they wish to access, which are often found on the Internet. Proxy, by definition, is a go-between or mediator that acts as one between a public network and a private network. After assessing client requests, the proxy server routes them to the relevant server if specific requirements are met. Proxies come in a variety of forms, some of which you should be familiar with for the test:

IP proxy: NAT secures networks by keeping machines anonymous. A router acts as an IP proxy, protecting clients on the LAN. IP proxies can be vulnerable to network attacks, so they should be regularly updated and monitored according to organization policies.

Caching proxy: Caching proxies, such as HTTP and WinGate, cache web pages from internet servers for a set time to save bandwidth and increase speed. They store or cache information for a 24-hour time limit to ensure no changes have occurred since the last request. Web browsers use a proxy auto-configuration function, which can be exploited and modified, redirecting users to malicious websites. Disabling these functions and auto-configuration in general within client web browsers can help prevent these risks. For example, if a co-worker accesses Google's website, the proxy caches the information for subsequent requests.

Proxy servers can perform various tasks such as applying policies, blocking websites, auditing employee usage, and scanning for malware. They can also protect a DMZ server's identity and provide secure tasks. Proxy servers typically have multiple network adapters, which should be periodically monitored for traffic and potential network attacks. However, they can be used maliciously to record unencrypted traffic, which could pose a security risk. Chaining multiple proxies can help mitigate this risk.

A proxy server can function as either a forward proxy, allowing clients to access websites or files via FTP, or a reverse proxy, where multiple HTTP or FTP servers use a proxy server to send content to clients. A reverse proxy can act as a layer of defense for an organization's servers. An application proxy, like Microsoft's Web Application Proxy, can act as a reverse proxy, allowing remote users to access multiple servers.

A proxy modifies client computer requests for anonymity, while transparent proxy provides increased efficiency but less protection for the client system, allowing for more flexibility in proxy usage.

Internet content filtering is a software application that filters out various Internet activities, such as websites, emails, and instant messaging. It functions as a content inspection device, disabling access to inappropriate content or websites that consume too much bandwidth. These filters can be installed on individual clients or as a proxy, securing the network by forbidding access to potentially malicious websites and blocking offensive material.

Internet filtering appliances analyze various data, including internet content, URLs, HTML tags, metadata, and security certificates. Some can even inspect malware. Web security gateways, like Forcepoint, scan for viruses, filter content, and prevent data loss. They actively monitor users' data streams to detect malicious code, bad behavior, or confidential data that should not be leaked outside the network.

There are numerous security device options available from various vendors, including all-in-one security appliances or unified threat management (UTM) devices, which can perform individual tasks or combine various features.

Note: Servers that more likely confront the Internet directly include proxies, content filters, and online security gateways. Prior to installation, security measures are needed for these "Internet-facing servers." Maintaining application updates and reviewing and implementing vendor-provided hardening documentation are the two most crucial security procedures. Before placing the proxy server—or any other server that faces the Internet—into a live environment, don't forget to do these steps.

Honeypots and Honeynets

Quick Hits!

Honeypots and honeynets trap potential attackers in monitored areas, containing dummy resources. These methods can be studied and analyzed, enhancing network security by analyzing the attacker's methods.

A honeypot is a single computer or unused IP address space, while a honeynet is a group of computers or servers. These devices usually store company information. A honeyfarm, consisting of various tools and a centralized group of honeypots, studies attackers.

A spam honeypot is a network administrator's tool that lures spammers into their network, allowing them to study their techniques and habits. This helps protect email servers, SMTP relays, and proxies over time. The information gained from studying spammers is often shared with other network admins or organizations to reduce spam.

Honeypots and honeynets pose risks to legitimate networks, requiring careful separation from them. They are often part of network intrusion detection systems, focusing on data loss prevention.

Data Loss Prevention (DLP)

Quick Hits!

Data loss prevention (DLP) systems, also known as data leak prevention devices, information leak prevention devices, or extrusion prevention systems, are designed to prevent the leakage of confidential data, particularly in communications, from a computer system or network.

Network-based Data Loss Prevention (DLP) systems detect and prevent confidential data from leaving the network, either hardware-based or software-based. Examples include detecting and preventing email transfer outside the network. Organizations like Check Point offer DLP solutions, and there are free open source applications. Cloud-based DLP is crucial for organizations with large BYOD or CYOD populations. Key elements of the security mindset include mitigation of security risks, understanding the cloud provider, in-depth analysis of code, and strong authentication, auditing, and logging. Proper planning can build confidence in the cloud, leading to a smoother transition and reduced risk.

Regarding DLP, surveillance of potentially exposed data may give rise to privacy issues.

It's crucial to study your company's privacy regulations before putting such a system in place. Poor DLP system installation can still result in leaks, thus it's critical to determine what kind of DLP solution your company requires, how exactly it will be implemented, and how it will be monitored.

NIDS vs. NIPS

Very Important!

You should be able to distinguish between a network intrusion prevention system (NIPS) and an intrusion detection system (NIDS) on the test; it's not a battle royale. We previously covered host-based intrusion detection systems (also known as HIDSs) in Chapter 4, "OS Hardening and Virtualization." Even while a lot of assaults can damage a single machine, a lot of network attacks also have the potential to bring down a server, switch, router, or even the whole network.

Network-based IPSs were created in an effort to stop these harmful network assaults, while network-based IDSs were created to identify them.

NIDS

By definition, a network intrusion detection system (NIDS) is a kind of intrusion detection system (IDS) that continuously monitors network traffic in an effort to identify hostile network activity, such as port scans and DoS assaults. NIDS is a tool for detecting rogue devices, access points, and network sniffers. Examples include Snort and Bro. Located at the network's entrance, NIDS should be fast and not a firewall. Planning the solution, device, and network connections is crucial to prevent network performance degradation.

This is how a Network Information Security Device (NIDS) can be implemented on a network, typically placed in front of a firewall. NIDS detects attacks and alerts administrators, while firewalls prevent them. Multiple NIDS points can be strategically placed around the network. Network administrators should monitor traffic using a network adapter configured for promiscuous mode.

A Network Information Security System (NIDS) can be used to reduce network costs by eliminating the need for HIDS on individual computers. However, it may not read encrypted packets or detect individual computer issues. To secure networks, organizations often combine NIDS and HIDS, monitor and update them regularly, and use Active Intrusion Detection Systems (NIIPS) for protection against attacks.

NIPS

A network intrusion prevention system (NIPS) is intended to examine network traffic and, in accordance with its configuration or security policy, eliminate, hold, or reroute any harmful traffic that it detects. NIPS and NIDS are application-aware devices that can interpret packets, determine their application, and allow or deny traffic on the network. Companies like Checkpoint and Snort offer NIPS solutions, which can remove or redirect malicious traffic and redirect attackers to a padded cell, ensuring no valuable information is lost.

A Network Information System (NIS) sits inline on a network, often in front of the firewall. It can be perimeter solutions or network core devices, with up to 16 ports for detection. These devices inspect packets for possible attacks, and need to be updated frequently to avoid misidentification of legitimate traffic or attacks.

False negatives occur when an IPS system fails to detect an attack's signature, allowing it to be mistaken for legitimate traffic. IPS systems can monitor for attack signatures and anomalies. Subversion errors can occur when NIDS/NIPS is altered by attackers, allowing false negatives and allowing attacks to enter networks. Protecting against these risks requires devices with IP masking capabilities and secure management console interfaces.

Newer NIPS solutions can act as protocol analyzers, protecting non-computer-based network devices like switches and routers. However, they are considered a single point of failure, making it necessary for organizations to install bypass switches for maintenance.

The NIPS policy should be implemented as either fail-close or fail-open, depending on the situation. Fail-close stops data transfer, while fail-open allows data transfer, including potential attacks, to pass through. For example, if a NIPS protects an individual server, a fail-close scenario would disconnect the system, causing risk for organizations requiring 100% uptime. In such cases, defense in depth is often the best strategy, such as firewalls filtering specific traffic but IPS filtering the bulk.

The Difference between NIDS and NIPS

Frequently asked!

Type of System	Summary	Disadvantage/Advantage	Example
NIDS	detects malevolent activity on networks	• Pro: A network only needs a certain number of NIDSs. • Con: It only identifies malevolent behavior.	• Snort • Bro IDS
NIPS	finds, eliminates, holds, and reroutes traffic	• Advantage: Identifies and reduces harmful behavior. • Pro: Has the ability to assess protocols. • Cons: Requires more energy. • Con: There might be false negatives and positives.	• Check Point • Systems solutions

The Protocol Analyzer's Role in NIDS and NIPS

Frequently asked!

Protocol analyzers like Wireshark and Network Monitor capture packets for analysis. However, some NIDS/NIPS solutions are full protocol analyzers, decoding application layer protocols like HTTP, FTP, or SMTP and forwarding results to an IDS or IPS analysis engine. This analysis can block multiple exploits based on a single signature, superior to basic pattern recognition. However, these solutions require more processing power and may be more expensive.

Note: WIDS and WIPS are wireless variants of IDS that are also available. They scan the radio spectrum for rogue access points and illegal access. Nonetheless, certain organizations may combine these terms within the notion of NIDS and NIPS. For both wired and wireless connections, be sure to utilize an IDS (or IPS).

Unified Threat Management

Frequently asked!

Unified Threat Management (UTM) is a new concept that combines various levels of defense into one solution, often referred to as a next-generation firewall (NGFW). Companies like Cisco, Fortinet, and Sophos offer UTM and NGFW solutions, often as a single device with a straightforward web-based GUI. However, a single point of defense can lead to a single point of failure, so secondary and backup UTM devices and server-based HIDSs provide a balance of depth and simplicity. UTMs should be quick to replace other devices, as their data processing and traffic flow requirements may be steep.

To ensure that you know how to use each, it was crucial to cover each tool and technology separately in this chapter. However, remember that a lot of these technologies are combined into a single solution, a tendency that is probably going to continue in the future.

CHAPTER 9:

ACRONYMS	LIST OF ABBREVIATIONS
PPTP	*Point-to-Point Tunneling Protocol*
WEP	*Wired Equivalent Privacy*
WAP	*Wireless Application Protocol*
TKIP	*Temporal Key Integrity Protocol*
WTLS	*Wireless Transport Layer Security*
NIST	*National Institute of Standards and Technology*
EMI	*Electromagnetic Interference*
RFI	*Radio Frequency Interference*
RATs	*Remote Access Trojans*
NIPS	*Network Intrusion Prevention Systems*
SSH	*Secure Shell*
STP	*Shielded Twisted Pair*
POTS	*Plain Old Telephone Service*
IDF	*Intermediate Distribution Frame*
MDF	*Main Distribution Frame*
PDS	*Protected Distribution System*
WAP	*Wireless Access Point*
SSID	*Service Set Identifier*
VPN	*Virtual Private Network*
WPS	*Wi-Fi Protected Setup*
AES	*Advanced Encryption Standard*
WTLS	*Wireless Transport Layer Security*
WPA	*Wi-Fi Protected Access*
WPA2	*Wi-Fi Protected Access 2*
TKIP	*Temporal Key Integrity Protocol*
CCMP	*Counter Mode with Cipher Block Chaining Message Authentication Code Protocol*
WEP	*Wired Equivalent Privacy*
WTLS	*Wireless Transport Layer Security*
PPTP	*Point-to-Point Tunneling Protocol*

Chapter 9

Wired Network and Device Protection

- **Protecting Wired Networks and Devices:** This part covers ways to lessen the possibility of an attack on your centrally located connected devices that manage network access as well as your wired networks. Topics addressed include coaxial, fiber-optic, and twisted-pair cable security as well as security for standard network equipment including SOHO routers and firewalls.
- **Securing Wireless Networks**: In this section, we'll discuss how to safeguard your wireless access points and keep your network safe from hackers, whether they are on the inside or outside of the building. Wi-Fi security, Bluetooth security, and security of other over-the-air technologies like RFID are among the wireless principles discussed.

As the security officer for your organization's wired and wireless networks, consider the type of cables used, vulnerabilities, future cabling plans, wireless network security, device access, and any outdated or outdated devices that need updating or removal.

This chapter discusses ensuring the security of network devices, cables, and mediums, starting with wired networks and cables, and then moving to wireless transmissions. It concludes with a network documentation summarizing security precautions from Chapters 6 through 9.

Securing Wired Networks and Devices

Implementing a security plan for your wired network is crucial. Chapters 6 and 7 discuss network design elements, protocols, threats, and firewalls. The wired network connects all these elements, making it targets for both internal and external attacks. Common vulnerabilities to network devices are discussed, along with potential threats to the network.

Network Device Vulnerabilities

Network devices like switches, routers, and fi rewalls can be vulnerable due to default accounts and weak passwords. Users can escalate privileges to access resources. Backdoors can be exploited by programmers. A good network administrator can protect against network attacks like denial-of-service, ensuring risk mitigation.

Default Accounts

Networking devices often come with a default user account, such as administrator or admin. To protect against attackers, it's important to change this account or remove it entirely. For instance, the SOHO router should be set up with a blank password and a complex password. Many devices also allow a separate user account with varying access levels, which should be set with a complex password or disabled.

Disable guest accounts on network devices and computers, and ensure complex passwords are used for guest accounts, as they are usually not enabled.

Weak Passwords

Passwords should be complex to prevent cracking by attackers. Strong passwords are crucial for protecting user accounts and network devices. A strong password should have at least eight characters, including uppercase letters, numbers, and special characters. The best passwords are 15 characters or more. Password checker programs like Password Meter and Kaspersky's can help identify weak and strong passwords.

Weak, Strong, and Stronger Passwords

Password	Strength of Password
Kyle	Weak
KyleBrown	Medium
Kyl7Own6	Strong
Iwant1@Ky9	Very strong (also known as "best")

Note: Password verification software may gradually modify its requirements. Furthermore, never use a password that has been input into a checker! The password strength checker's objective is to illustrate the strength of a password so that you can determine what standards to use to future passwords.

Table 9-1 outlines three password types: weak, medium strength, and strong. The weak password has only 6 characters, the medium password has 11 characters and 2 uppercase characters, and the strong password uses the pipe symbol. The strong password takes years to crack due to the increasing power and efficiency of computers. Additionally, a botnet can brute-force a password faster than one system, making it crucial for security admins with high-performing employees.

Privilege Escalation

Privilege escalation is when an attacker exploits a bug in an operating system or application to gain protected access to resources. This technique is used by attackers to bypass digital rights management (DRM) on games and music, using methods like jailbreaking. Patches are released if issues are discovered, and malware also attempts to exploit privilege escalation vulnerabilities. Network devices can also be used for privilege escalation, as long as they are regularly updated and checked for updates. Your network may be in for some "fun" if you combine this with the fact that someone neglected to set a password for the user account or disabled it. Some devices also provide the option to encrypt the firmware. Here are a few examples of distinct privilege escalation scenarios:

- **Vertical Privilege Escalation:** Privilege elevation occurs when a lower privileged user accesses functions reserved for higher privileged users, such as an administrator. To protect against this, update network device firmware, operating systems, and use access control systems like UAC.

- **Horizontal Privilege Escalation:** Normal users can access reserved functions or content through hacking or by reading others' emails. It's crucial to lock computers or log off when not physically present.

Privilege de-escalation allows high-privileged users to downgrade access, potentially allowing sneaky admins to gain confidential information. Network security requires a three-dimensional approach.

Back Doors

A backdoor is a means of obtaining unauthorized remote access to a computer, getting around standard system authentication, or breaking into a cryptosystem by breaking the rules.

Backdoors, initially used by developers as legitimate access to applications, have since become a popular method for attackers. They can be programs like RATs or rootkits. Nowadays, they are less common due to discouragement by software manufacturers and network device makers.

Network Attacks

Network attacks like denial-of-service can cause significant damage to network devices. To prevent these attacks, keep up-to-date systems and devices, and use smart network intrusion prevention systems (NIPSs).

Other Network Device Considerations

Network administrators often use remote ports on devices for remote administration, which can be used maliciously. To protect these devices, disable them, use a strong authentication system, and consider data encryption.

Network administrators often use Telnet to access network equipment remotely, but this practice is not secure due to its lack of encryption, authentication, and associated vulnerabilities. Instead, administrators should use Secure Shell (SSH) protocol.

Cable Media Vulnerabilities

Cabling, including cables, network jacks, patch panels, and punch blocks, is often overlooked in a network. It's crucial to consider the types of cabling, their vulnerabilities, and strategies to mitigate risk.

- **Twisted-pair:** Twisted-pair cables, commonly used in network connections, consist of four pairs of wires twisted together, sending electrical signals for data transfer. However, crosstalk is a common security issue.
- **Fiber-optic:** Fiber-optic cables are glass/plastic-based cables that transmit data using light instead of electricity. They are secure, resistant to EMI, RFI, and wiretapping. They are used in high-bandwidth networking technologies like Fibre Channel (FCoE and FCIP) with single-mode and multimode connectors, and are used in single-mode and multimode applications.
- **Coaxial:** Coaxial cable, with a single copper core, is less commonly used for internet or video connections, often used in smaller companies' networks. Data emanation is a common security risk.

Let's discuss a few of the intrinsic weaknesses that each of these cables possesses right now.

Interference

Anything that tampers with or alters a signal as it travels across a wire is considered interference. Although there are several forms of interference, you need be familiar with the following ones in order to pass the test:

- **Electromagnetic interference (EMI):** Electromagnetic interference (EMI) can affect electrical circuits, devices, and cables, including TVs, microwaves, air-conditioning units, motors, and unshielded lines. Copper-based cables and network devices should be avoided, and shielded cables can be used if possible. Electrical cables should be BX (encased in metal) and not Romex (not encased in metal) to meet building codes. Radio jamming can also be a concern.
- **Radio frequency interference (RFI):** Interference from AM/FM transmissions and cell towers, part of the EMI family, can be a significant threat to businesses. To defend against RFI, methods like EMI can be employed, and filters can be installed to eliminate signal frequency. Wireless signals can also interfere with devices, so it's important to keep them away.

Crosstalk is another frequent form of interference that will be covered next.

Crosstalk

Crosstalk occurs when a signal on one copper wire causes an unwanted effect on another, causing a "bleeding" effect. This occurs when telephone lines are close together, causing intermittent signal jumps. It can also occur with modem connections, affecting data transfer.

Note: Crosstalk on cell phones is often referred to as co-channel interference, or CCI.

Twisted-pair cable is a method to prevent crosstalk in analog signals sent across wires, particularly in standard POTS connections and computer modems. The twists reduce the risk of signal leakage to other wires. Digital signals like Ethernet data transfers or Voice over IP are less susceptible to crosstalk. Crosstalk can occur due to tightly bundled cables, which can cause damage. Twisted-pair cabling can be divided into near end and far end crosstalk.

Shielded twisted-pair (STP) can be used to minimize crosstalk in twisted-pair cables, a common method used for digital data transmissions, when interference is too high for effective data transmission.

Data Emanation

Data emanation is the electromagnetic field generated by a network cable or device, which can be manipulated to eavesdrop on conversations or steal data. It is a common security risk when using coaxial cables and other copper-based cables. Solutions to mitigate this include using shielded cabling, shielding devices, and protecting server rooms and cabling, especially if the data being transmitted is confidential. Studies are ongoing to understand signal emanations and contain them, with TEMPEST standards being used to investigate emissions from electrical and mechanical devices.

Tapping into Data and Conversations

Any system can be accessed or hacked, depending on network, cabling, and security precautions, with the extent of the breach varying depending on the extent of the breach.

Taping, originally used by attackers to physically bite into wires for network monitoring, now encompasses various meanings, including wiretapping, which can involve various actions.

- **Connecting to a punch block, or RJ11 jack with a butt set:** A butt set is a phone-like device with alligator clips that connects to phone equipment terminals, allowing users to test lines or listen to conversations. It's primarily used by telecommunications technicians but can be used for malicious purposes. To protect against this, keep phone equipment in a secure room or wiring closet. Installing a network of cameras is cheaper.
- **Plugging in to an open port of a twisted-pair network:** Disable unused ports on RJ45 wall plates or open ports in server rooms, lock central connecting devices, and restrict access to only authenticated individuals in the server room.
- **Splitting the wires of a twisted-pair connection:** Cutting twisted-pair cables and soldering a second cable to wires can disrupt communications for individual computers or segments. Cables should not be exposed, and runs should be above the drop ceiling and inside walls. Patch cables are necessary for connecting to RJ45 jacks, but an alert can be sent to junior administrators.
- **Using a spectral analyzer to access data emanations:** Spectral analyzers measure electrical waveform composition and decode encrypted

transmissions. They should be used only by authorized personnel in buildings. Metal detectors and video cameras can detect and prevent mischievous attempts to decode transmissions.

- **Using a passive optical splitter for fiber-optic networks:** The expensive and difficult-to-implement optical splitter could cause chromatic dispersion on the fiber segment, potentially leading to data loss. Administrators can monitor the network for dispersion and receive alerts if it occurs. Chromatic dispersion, like the rainbow, can occur when light is refracted into a rainbow of colors, potentially causing data transmissions to fail.

Commercial office buildings typically have two types of wiring frameworks: intermediate distribution frame (IDF) rooms and main distribution frame (MDF) rooms. These rooms are vulnerable attack points, with MDF rooms being better protected than IDF rooms. Organizations are increasing security on both types, treating them like data centers or high-end server rooms. Security admins should consider methods like multifactor authentication, biometrics, and active surveillance systems. An attacker can exploit vulnerabilities from within and without by connecting to SNMP-monitored systems.

Some organizations and government agencies mandate the use of a protected distribution system (PDS) for wired connections, securing unencrypted transmission of classified information through cables, terminals, and equipment.

The Security+ exam covers wiretapping, a common hacking technique, but not in-depth. Understanding the basics of wireless networks is crucial for a secure network.

Securing Wireless Networks

Key Topic!

Wireless networks pose new challenges for network administrators, requiring secure wireless access points and adapters. This section covers securing WAPs, adapters, and other wireless devices, highlighting vulnerabilities like Wi-Fi, Bluetooth, and RFID.

The Administration Interface

To secure WAPs, first check the administration interface or console, which is typically a simple and weak password. Modify the password to a complex one and consider disabling remote administration if not necessary. If enabling it, ensure the password is complex.

> ***NOTE***
>
> ***Wireless Access Point Vulnerabilities***
>
> The wireless access point (WAP) is the central device for wireless network adapters, requiring secure broadcasts, strong encryption, and monitoring for rogue access points and network nomads.

SSID Broadcast

Quick Hits!

The Service Set Identifier (SSID) is a broadcast used by Wireless Access Points (WAPs) to identify the network and the name used by clients to connect to the wireless network. Disabling the SSID broadcast provides a preliminary level of security, preventing attackers from accessing the network. Temporarily enabling the SSID and reducing transmitter power can help secure the area the WAP serves.

Note: Clients can connect to a wireless network without broadcasting the SSID manually within third-party wireless adapter settings or in Windows by starting WLAN AutoConfig service and accessing the Managing Wireless Networks link. However, they must know the SSID, encryption type, and encryption key. If disabled, new wireless clients cannot connect.

Rogue Access Points

Quick Hits!

Rogue APs are unauthorized wireless access points that allow access to secure networks. Companies should monitor their network with network documentation and use network mapping programs and Microsoft Visio to detect and document rogue APs. Older WAPs should be updated, disabled, or disconnected. Using a network mapping program, Windows' wireless application, or third-party applications can help. NIDS or NIPS solutions can help detect and prevent rogue devices.

Evil Twin

Always Remember!

An evil twin is a counterfeit, unauthorized wireless access point that uses the same SSID name as a nearby network. It is controlled by a person with phishing motives, potentially gaining valuable information and passwords from users. To protect against this, virtual private networks (VPNs) can be implemented, administrators should scan networks for rogue APs, and users should be trained not to send sensitive information over wireless networks.

Weak Encryption

A must know!

Weak or no encryption can damage a wireless network, especially if connecting older or outdated devices. Strong encryption is crucial, and Wi-Fi Protected Access 2 (WPA2) is the best protocol. Ensure WAP and network adapters have the same encryption level. Table 9-2 lists available wireless protocols and encryption types.

Wireless Protocol	Description	Encryption Level (Key Size)
WEP	• Wired Equivalent Privacy (Deprecated)	• 64-bit • Also 128-bit but uncommon
WPA	• Wi-Fi Protected Access	• 128-bit
WPA2	• Wi-Fi Protected Access Version 2	• 256-bit
TKIP	• Temporal Key Integrity Protocol • (Deprecated) Encryption protocol used with WEP and WPA	• 128-bit
CCMP	• Counter Mode with Cipher Block • Chaining Message Authentication • Code (CBC-MAC) Protocol • Encryption protocol used with WPA2	• 128-bit

	• Addresses the vulnerabilities of TKIP • Meets requirements of IEEE 802.11i	
AES	• Advanced Encryption Standard • Encryption protocol used with WPA/WPA2 • Strongest encryption method in this table	• 128-bit, 192-bit, and 256-bit
WTLS	• Wireless Transport Layer Security	• Based on TLS

WEP is the weakest type of encryption, followed by WPA and WPA2. It's better to have WEP, use complex encryption keys, and change them frequently. Some devices can support WPA through firmware upgrades or software add-ons. A typical WAP with WPA2 and AES is shown. A WPA2-PSK is used for WPA2 protocol, not RADIUS.

The Wireless Transport Layer Security (WTLS) protocol, part of the Wireless Application Protocol (WAP) stack, uses encryption and hashing algorithms for secure user sessions.

Wi-Fi Protected Setup

Wi-Fi Protected Setup (WPS) is a security vulnerability, created to enable easy connectivity to wireless access points. Its eight-digit code can be broken by brute-force attacks. To prevent this, disabling WPS in the firmware interface or upgrading to a newer device is recommended.

Ad Hoc Networks

Ad hoc networks, which allow wireless connections without a centralized WAP or WLAN controller, are inherently insecure and should be disabled by security admins on various devices.

VPN over Open Wireless

VPN connections are secure sessions through an encrypted tunnel, best in wired environments. However, sometimes a wireless VPN connection is necessary. VPN over open wireless is inherently insecure and should be scanned for. Encryption is necessary, such as Point-to-Point Tunneling Protocol (PPTP), IPsec, or another secure protocol. In a standard Cisco wireless VPN configuration, the wireless client initiates a tunnel to a VPN server via a wireless LAN controller. In smaller offices or home offices, a single device can act as an all-in-one solution.

Wireless Access Point Security Strategies

The optimal location for a Wireless Access Point (WAP) is in the center of the building, ensuring equal access and minimizing signal interference. To reduce transmission power and avoid interference, WAPs should be kept away from electrical panels, cables, devices, or motors. If necessary, shield the device or access point itself. To secure wireless connections, understand different wireless systems and antenna types. Point-to-multipoint systems, such as

SOHO wireless routers, connect multiple devices in any direction using omnidirectional antennas. A router may have multiple vertical omnidirectional antennas, such as three MIMO antennas for higher data throughput. Positioning antennas parallel or at an angle can improve wireless transmission coverage. For example, if you have 100 computers on one floor and two WAPs, position them at vertical angles and position them in a 90-degree sweep for excellent coverage.

Note: Regarding 802.11ac and 802.11n, channel and channel width choices can also affect security and performance. Generally speaking, 5-GHz frequency bands work better than 2.4-GHz bands, although attackers may watch both over the air. While auto-configuring on certain WAPs can be quite effective in finding the least occupied wireless frequencies and channels, it can also pose a security concern. Think about utilizing fewer frequently used channel numbers and maybe a smaller channel, such as 40 MHz for 802.11ac rather than 80 MHz. Moreover, keep in mind to minimize antenna power levels. Once more, to get the highest level of organizational efficiency, you must thoroughly test and then balance security and performance.

Wireless site surveys are essential for wireless systems, as they collect WLAN data and signal information. There are three types: passive, active, and predictive. Passive surveys listen to WLAN traffic, measure signal strength, while active surveys measure data transfer rate and packet loss. Predictive surveys simulate real-world data, such as the WAP and computer distance.

Surveys can identify nonaggressive interference, such as neighboring radio waves and electrical equipment, and aggressive jamming techniques like wireless signal jammers. These jammers can initiate denial-of-service attacks by creating random noise or using powerful radio signals. Wireless software like NetStumbler can locate and remove jammers, as well as identify potential wireless replay attacks.

Wireless Access Points (WAPs) often have built-in firewalls, which should be enabled for stateful packet inspection (SPI) and NAT filtering. WAPs can also be configured for MAC filtering, limiting access to network adapters with corresponding MAC addresses. If a device broadcasts this MAC table, update the firmware and adjust broadcast range. Strong encryption and other network access control methods, such as 802.1X and RADIUS, are crucial for security.

WAPs can support isolation mode, preventing communication between clients while allowing access to the Internet or other networks. IEEE 802.1X standard port-based network access control can be included, but a WAP must also act as a router, serving as a wireless gateway.

Encryption technologies like SSL, SSH, or PGP can secure data transmissions from attackers. However, authentication and a strong wireless protocol like WPA2 with AES are the best security precautions to protect against network attacks.

Wireless LAN controllers are a smart solution for organizations with multiple Wireless Access Points (WAPs). These controllers act as switches, increasing data transfer speeds and centralizing security management. They are considered "thin access points" due to reduced functionality, unlike fat access points that contain all functionality. WLAN controllers are used by most midsized companies and enterprise networks. Some companies opt to not run wireless due to costs, administration, maintenance, and security. However, it is crucial to disable wireless functions on devices that enable wireless access, including wireless adapters and Bluetooth transmitters.

Wireless Transmission Vulnerabilities

Wireless networks transmit information through air or space, making data easily identifiable. War-driving is an attack on wireless networks, where individuals search for networks using devices with strong antennas. They crack passwords or passphrases to gain access, with over 40% of networks unprotected. Password cracking can be done by guessing or brute-force attacks. More information on password cracking can be found in Chapter 12.

To protect against war-driving, organizations should hide the WAP's SSID, position it properly, reduce power levels, use strong encryption, and change the passphrase regularly. War-chalking, where symbols are drawn in public places to denote open, closed, or protected wireless networks, can also be used by organizations to inform the public about open networks. Professional presentation is essential.

IV attacks are vulnerabilities in wireless networks, where an attacker observes a cipher's operation using multiple keys and deciphers data. The small initialization vector in WEP or WPA packets, known for its insecurity, is a significant issue. Stronger protocols like WPA2 and AES can help prevent IV attacks.

DoS attacks can be used against Wireless Access Points (WAPs) in various ways, such as using spoofed MAC addresses or Wi-Fi disassociation attacks. Incomplete authentication can lead to the loss of session spots, which can be prevented by setting expiration timeouts, updating the WAP, and implementing wireless frame protection. Wi-Fi disassociation attacks target users connected to a Wi-Fi network, forcing them to de-authenticate and re-authenticate to the WAP. Automated software tools can help identify the WAP's SSID or ESSID, and the WPA passphrase using a dictionary attack. These attacks can be prevented by using WPA2, complex passphrases, and management frame protection.

WAPs can be vulnerable to brute-force attacks and exhaustive key searches, which can be prevented by limiting password attempts, using time delays, and requiring complex answers during authentication, and blacklisting attacker IP addresses.

Bluetooth and Other Wireless Technology Vulnerabilities

NFC, along with Bluetooth, RFID, and other wireless networking technologies, can technically be considered a wireless network if two or more wireless devices communicate with each other.

Bluetooth is susceptible to attack, including bluejacking and bluesnapping. It can also be conflated with other wireless technologies, such as WLAN or Wi-Fi standards. To avoid these issues, use Bluetooth version 1.2 devices or higher, which use adaptive frequency hopping. Place Bluetooth and WLAN access points in different areas of the building. Companies may have policies governing Bluetooth usage, limiting its use outside buildings or using 5-GHz WLAN standards.

Near field communication (NFC) allows Bluetooth-equipped devices to pair and transmit data over short distances. However, NFC is not secure, as data can be destroyed by jammers and users risk replay attacks. To prevent these risks, users should use secure applications with SSL/TLS or other channels during NFC sessions.

Bluejacking

Bluejacking is the sending of unsolicited messages to Bluetooth-enabled devices, often appearing malfunctioning. It can include images and sounds and is used in less-reputable marketing campaigns. Stopping bluejacking can be achieved by setting the device to "undiscoverable."

Bluesnapping

Bluesnapping is the unauthorized access of information from a wireless device through a Bluetooth connection, typically involving data theft. To prevent this, pairings must be made between the attacker and the victim. Discouragement methods include using a unique pairing key, setting Bluetooth devices to "undiscoverable" or turning off Bluetooth altogether, especially in playgrounds like Times Square. Differentiating between bluejacking and bluesnapping is crucial for exam preparation.

RFID and NFC

Although radio-frequency identification (RFID) has various applications, its fundamental function is the identification and tracking of tags affixed to objects. In the context of security, it usually refers to authentication.

RFID technology is vulnerable to attacks like skimming, MIMO, sniffing, eavesdropping, spoofing, and jamming. Attackers use these to steal passcodes and reverse engineer tags. To prevent these attacks, security admins should consider newer devices, encryption, chip coatings, data filtering, and multifactor authentication methods. Encryption, rolling codes, and challenge-response authentication are effective prevention methods.

RFID ranges can range from 10 cm to 200 meters, potentially allowing information skimming by attackers. To avoid this, use newer RFID proximity readers and keys from trusted companies. Near field communication (NFC) requires devices to be within 4 cm of each other, making skimming difficult. NFC benefits contactless payment systems.

More Wireless Technologies

Many companies deny cellular access, relying instead on Wi-Fi methods. This is common in choose your own device (CYOD) and corporate-owned, personally enabled (COPE) environments. If cellular is necessary, security can be increased by using newer devices, updating OS, PRL, identification technologies, and VPNs. Mobile employees should understand the risks of open Wi-Fi networks.

GPS usage should be carefully evaluated, considering its benefits against potential vulnerabilities. Disable GPS whenever possible and consider disabling other GPS and geolocation-related technologies like geotagging and geofence. Geotagging can be useful in social media but can be exploited by attackers. Geofence, a virtual fence, can alert users to physical premises and provide security for wireless networks by defining physical borders.

Organizations using satellite communications (SATCOM) face risks from malware if their firmware is not updated. Updating may require physical access to the parabolic antenna, increasing the risk of hacked devices. Secure planning is necessary for physical access, firewalling, and updating. Wireless technologies are constantly evolving, making it crucial to be vigilant and prepare a defense against potential hacks.

CHAPTER 10:

ACRONYMS	LIST OF ABBREVIATIONS
CHAP	*Challenge-Handshake Authentication Protocol*
RAS	*Reliability, availability, and serviceability*
MD5	*Message Digest 5*
MPLS	*Multiprotocol label switching*
GRE	*Generic Routing Encapsulation.*
MFA	*Multifactor Authentication*
CCTV	*Closed-circuit Television*
IP	*Internet Protocol*
3DES	*Triple Data Encryption Standard*
PIV	*Personal Identity Verification*
CAC	*Common Access Card*
SSO	*Single Sign-On*
FIdM	*Federated Identity Management*
SAML	*Security Assertion Markup Language*
LDAP	*Lightweight Directory Access Protocol*
LDAPS	*Secure Lightweight Directory Access Protocol*
IEEE	*Institute of Electrical and Electronics Engineers*
PNAC	*PNAC - Port-based Network Access Control*
EAP	*EAP - Extensible Authentication Protocol*
TLS	*TLS - Transport Layer Security*
VPN	*VPN - Virtual Private Network*
RAS	*RAS - Remote Access Service*
MS-CHAPv2	*Microsoft Challenge-Handshake Authentication Protocol version 2*
SPAP	*Shiva Password Authentication Protocol*
PPTP	*Point-to-Point Tunneling Protocol*
L2TP	*Layer 2 Tunneling Protocol*
RRAS	*Routing and Remote Access Service*
GRE	*Generic Routing Encapsulation*
AAA	*Authentication, Authorization, and Accounting*
UDP	*User Datagram Protocol*
TCP	*Transmission Control Protocol*
SCTP	*Stream Control Transmission Protocol*

Chapter 10

Physical Security & Authentication Models.

Key Topic!

Life is about proving oneself, and security is no exception. To gain access to an organization's building and resources, one must first prove themselves physically, providing indisputable evidence of their identity. Authentication is when a person's identity is confirmed or verified through a specific system, and authorization is when a user is given permission to access certain resources. The CompTIA Security+ exam focuses on authentication and access control, with this chapter focusing on the authentication portion of access control. Authentication is required to gain access to secure areas of the building or secure data. People can identify themselves in various ways, such as knowledge factors, possession factors, signatures, inherence factors, and location. Identity proofing is an initial validation of an identity, such as when employees working for a government agency enter a restricted building and a guard checks their ID. It can also be used for anonymous access. As you read about physical and logical authentication technologies, remember this introduction and apply these concepts to each of these authentication types.

Physical Security

Physical security is the first line of defense for an organization, like a firewall. Implementing security methods should be a top priority, but it can sometimes slip to the bottom of the list. Proper building entrance and secure access to equipment are vital.

General Building and Server Room Security

Building security is crucial for an organization's overall security, as it reduces the need for an authentication system. Surveying the building's perimeter, scanning for hiding places, and maintaining well-lit areas are essential steps. Security guards and guard dogs should be trained properly, and video surveillance can be employed to track user movements. Cameras should be placed on the exterior perimeter, in hard-to-access areas, and inside secure areas like executive offices, wiring closets, server rooms, and research and development areas. Closed-circuit television (CCTV) or wired/wireless IP-based solutions can be used, but the video stream should not be broadcast. Motion detectors and other sensors are also common in total alarm systems.

Server room security is crucial for organizations, and it should be accessible only to authorized IT personnel. Devices and servers should have complex passwords and be physically locked down with cable locks. Network cabling should be approved and meet the requirements for protected distribution systems. Security also includes user safety, with well-planned fire drills, exit signs, escape plans, and escape routes. Secure door access is the primary way to impede unauthorized access, ensuring that intruders cannot be authenticated.

Door Access

Door locks are essential for businesses, and it is crucial to consider all types of security, including entrances to the building. Regular key locks should be tested and a backup should be in place in case of a break-in attempt. Cipher locks are a good solution when regular key locks are not enough but don't want to implement an electronic system. Electronic access control systems like cardkey systems are common, using scanning devices on each door for access. These cardkeys should be logged and protected by a cardkey controller. Some systems use photo ID badges for identification and authentication. Hardware-based security tokens are physical devices given to authorized users to help with authentication. They can be attached to a keychain or part of a card system.

Smart cards are used for various applications, such as securing computers and encrypting emails. Examples include the PIV card (Personal Identity Verification) and the Common Access Card (CAC). Advanced smart cards have specialized cryptographic hardware that can use algorithms such as RSA and 3DES. Smart card security systems usually consist of the smart card itself, smart card readers, and a back-office database.

Older technologies like proximity sensors are less secure than modern technology, but they can be more expensive and time-consuming. Companies like HID offer various levels of door access control systems. To increase security, organizations may implement mantraps, areas between two doorways, security guards, video surveillance, multifactor authentication, and sign-in logs. Biometrics can be added to door access systems to create a different type of multifactor authentication scheme.

Biometric Readers

Biometrics is the science of recognizing humans based on physical characteristics, such as fingerprints, retinal patterns, iris patterns, voice patterns, and facial/bone structure. It is used for authentication, access control, and identifying individuals under surveillance. Biometric readers are becoming more common in door access systems, laptops, and USB devices. Biometric information can also be incorporated into smart card technology. A security admin should monitor biometric systems for errors, such as false acceptance and false rejection, and analyze the crossover error rate (CER) to maintain a common error rate. Biometric systems have been easily compromised in the past, but recent models have started to live up to their hype. Thorough investigation and testing are necessary before purchase and installation, and a multifactor authentication scheme is recommended. Voice recognition software has also made significant strides, making a combination of biometrics, voice recognition, and PIN access an excellent three-factor authentication system.

Authentication Models and Components

Key Topic!

The text discusses the importance of selecting the right authentication model, technology, and monitoring system for operating systems and computer networks. It also discusses the necessity of local or remote authentication, and the use of Windows-based or third-party solutions. Examples of possible solutions are provided.

Authentication Models

Many small and midsized businesses use one type of authentication, such as username and password, for accessing computer networks. However, this is not enough due to the increasing complexity of passwords and the increasing power of password-cracking programs. Organizations sometimes use multiple factors of authentication, such as multifactor authentication (MFA), smart card, biometrics, and a username/password. However, these schemes can be static and require more information and identification.

Context-aware authentication, also known as context-sensitive access, is an adaptive way of authenticating users based on their usage of resources and the system's confidence in the user. Single sign-on (SSO) is a method where a user can log in once but gain access to multiple systems without being asked to log in again. It is meant to reduce password fatigue and IT help desk calls, while also reducing phishing.

SSO is a derivative of federated identity management (FIdM), where a user's identity and attributes are shared across multiple identity management systems. Examples of SSO systems include Shibboleth, which allows users to sign in with a single digital identity and connect to various systems run by federations of different organizations.

Web-based SSO can be problematic due to disparate proprietary technologies. To address this issue, the XML-based Security Assertion Markup Language (SAML) and the OpenID Connect protocol were developed. These protocols use REST/JSON message flows and specify separate roles for the user, service provider, and identity provider.

To ensure proper functioning of any authentication scheme, it is essential to monitor it periodically and block people who cannot provide proper identification.

Localized Authentication Technologies

Various authentication technologies, such as LDAP and Kerberos, are used to authenticate users to local area networks. These technologies are classified based on their usage, with encryption concepts and protocols detailed in Chapters 14 and 15, "PKI and Encryption Protocols."

802.1X and EAP

802.1X is an IEEE standard that defines port-based network access control (PNAC), a data link layer authentication technology used to connect hosts to LANs or WLANs. It allows security control that ties physical ports to end-device MAC addresses and prevents additional devices from being connected to the network. The process starts with a central connecting device such as a switch or wireless access point, which must first enable 802.1X connections and have the 802.1X protocol and supporting protocols installed. The client computer, known as the supplicant, needs an operating system or additional software that supports 802.1X.

802.1X encapsulates the Extensible Authentication Protocol (EAP) over wired or wireless connections, and it defines message formats. An example of an 802.1X-enabled network adapter is shown in Figure 10-1. The connection consists of three components: a supplicant, an authenticator, and an authentication server.

Several types of EAP authentication exist, including EAP-MD5, EAP-TLS, EAP-TTLS, EAP-FAST, and PEAP. Cisco also created a proprietary protocol called LEAP (Lightweight EAP), which requires a Cisco device and a third-party client on Windows computers to connect to the Cisco device.

802.1X can be used for port-based network access control on LANs, especially VLANs, and VPNs as a way of remote authentication. The protocol was originally defined in 2001 and has been redefined in 2004 and 2010.

LDAP

Key Topic!

The Lightweight Directory Access Protocol (LDAP) is a protocol used for accessing and modifying directory services data, part of the TCP/IP suite. It was initially used in WAN connections and has evolved into a common protocol for services like Microsoft Active Directory on Windows Server domain controllers. LDAP controls the directory service, organizing users, computers, and other objects within Active Directory. Windows servers running Active Directory use parameters and variables to query object names. Secure LDAP, also known as LDAPS, protects Active Directory from tampering, using SSL on top of LDAP.

Kerberos & Mutual Authentication

Key Topic!

Kerberos is an authentication protocol developed by MIT that enables secure two-way authentication between computers. It is commonly used in client-server environments, where both the client and server verify each other's identities. Kerberos protects network servers from illegitimate login attempts, similar to the three-headed guard dog Cerberus guarding Hades. A common implementation of Kerberos occurs when a user logs on to a Microsoft domain, where the key distribution center (KDC) works with tickets to prove user identity. The client computer attempts to authenticate to the authentication server portion of the KDC, receiving a ticket to get other tickets (TGT). The client uses this ticket to demonstrate its identity to a ticket-granting server, aiming to access a service. Kerberos is designed to protect against replay attacks and eavesdropping, but it relies on a centralized server, which can be a single point of failure. To mitigate this, secondary and tertiary domain controllers can be installed. Kerberos is vulnerable to attack, with older Windows operating systems vulnerable to privilege escalation attacks and newer Windows systems to spoofing.

Remote Desktop Services

Key Topic!

Remote Desktop Services (RDP) allows remote control of Windows computers from a client computer, enabling access to specific applications. It is bundled within Remote Desktop Management Server in Windows Server 2008 and via the Remote Desktop Services server role in Windows Server 2012 R2. RDP authenticates remote users if they have been properly configured. It integrates with standard Windows Server authentication and has an inbound port of 3389 for accepting connections. Client sessions are stored at the Windows Server running RDP, allowing for disconnections and reuse.

Vulnerabilities to RDP include a well-known port, weak encryption, and a lack of multifactor authentication. To protect data, organizations may choose to use another remote control application. Third-party applications like VNC and TeamViewer can be more secure, using web-based connections via HTTPS on port 443. RDP is generally limited to SSL (TLS 1.0) with a 128-bit key based on the RC4 algorithm. Captive portals, where users are redirected to a web page for authentication, can be circumvented with packet sniffers like

Wireshark. Extended or multifactor authentication can be used to avoid these risks.

Remote Authentication Technologies

Remote authentication technologies like RAS, VPN, RADIUS, TACACS+, and CHAP are crucial for preventing illegitimate connections and increasing the security of your network.

Remote Access Service

Remote Access Service (RAS) was initially designed to enable dial-up connections from remote clients, but it is still used as a fail-safe in many network operation centers and server rooms. To secure a RAS server, it is essential to deny access to individuals who don't require it and monitor daily logs. The Challenge-Handshake Authentication Protocol (CHAP) is a secure method for RAS authentication, using DES and MD5 encryption types. Microsoft developed its own version, MS-CHAPv2, which allows for mutual authentication between the client and authenticator. Other RAS authentication protocols include SPAP and PAP, which are insecure and should be avoided.

CHAP authentication involves a three-way handshake to verify the identity of the client, and the data transmitted in these RAS connections is encrypted. By default, Microsoft RAS connections are encrypted by the RSA RC4 algorithm. Dial-up connections are important for supporting protocols, authentication types, and encryption types used in other technologies, and some organizations still use them for remote users or administrative purposes. VPN connections using dial-up, cable Internet, and DSL are also considered remote access.

Virtual Private Networks

A virtual private network (VPN) is a secure connection between two or more computers or devices not on the same private network. VPNs use the Internet to connect one host to another, and data encapsulation and encryption are used to ensure only proper users and data sessions reach a VPN device. Tunnels are created through any LANs and WANs that might intervene, connecting the two VPN devices together. Common VPN connections include PPTP and L2TP, which can cause errors if security settings are not properly configured on the client and server sides.

Site-to-site VPNs are generally more secure because an admin can specify that only specific networks can connect in a private intranet fashion. Microsoft VPNs can be set up on standard Windows Server by configuring Routing and Remote Access Service (RRAS), and remote access policies can be created to permit or deny access to groups of users for dial-in or VPN connections.

Older VPNs use either PPTP (port 1723) or L2TP (port 1701) with IPsec, and can incorporate CHAP on the client side and RADIUS servers for authentication. Newer VPNs protect traffic by using SSL or TLS, such as OpenVPN. SSL/TLS solutions for VPN improve endpoint security and enable always-on VPN functionality.

Cisco systems use the Generic Routing Encapsulation (GRE) protocol to encapsulate routing information between VPN-enabled connected networks that use PPTP or IPsec. The Gover GRE feature provides a mechanism for tunneling MPLS packets over a non-MPLS network, creating a virtual point-to-point link across non-MPLS networks.

RADIUS vs. TACACS

Key Topic!

RADIUS is a client-server protocol that provides centralized administration of dial-up, VPN, and wireless authentication. It works within the AAA concept and checks if the correct authentication scheme is used when clients attempt to connect. RADIUS uses port 1812 for authentication messages and port 1813 for accounting messages, both using UDP as the transport mechanism. RADIUS federation is a concept where multiple RADIUS servers need to communicate safely, creating trust relationships and managing authentication requests. Diameter protocol is another AAA protocol similar to RADIUS, evolving from RADIUS by supporting TCP or SCTP but not UDP. Terminal Access Controller Access-Control System (TACACS) is another remote authentication protocol used more often in UNIX networks. TACACS+ is a more commonly used implementation of TACACS, using TCP as the transport mechanism instead of UDP. RADIUS encrypts only the password in the access-request packet, while TACACS+ encrypts the entire body of the access-request packet, encrypting entire client-server dialogues and providing more types of authentication requests than RADIUS.

CHAPTER 11:

ACRONYMS	LIST OF ABBREVIATIONS
DAC	*Discretionary Access Control*
ACL	*Access Control List*
MAC	*M Mandatory Access Control*
RBAC	*Role-Based Access Control*
ABAC	*Attribute-Based Access Control*
DAC	*Discretionary Access Control*
OU -	*Organizational Unit*
ADU	*Active Directory Users and Computers*
FIM	*Federated Identity Management*
ACE	*Access Control Entry*
DACL	*Discretionary Access Control List*
SACL	*System Access Control List*
R, W, X	*Read, Write, Execute*
UAC	*User Account Control*
OU	*Organizational Unit*
ADUC	*Active Directory Users and Computers*
FIM	*Federated Identity Management*
ACL	*Access Control List*

Chapter 11

Access Control Methods and Models

Controlling user access is crucial for maintaining security on a computer network. First, define who needs access and what they need to access. Develop an access control plan based on your organization's procedures, policies, security level, and IT resources. Implement safe practices to enhance the model's effectiveness. Implement secure users, groups, and network objects, including policies and Group Policy objects, to enforce adherence to organizational rules.

Access Control Models Defined

Access control models manage and organize access to physical areas and computer systems, crucial for users accessing secure data. Combining best practices like separation of duties, job rotation, and least privilege can create a robust plan for secure areas. There are several models with unique characteristics.

Discretionary Access Control

Discretionary access control (DAC) is an access control policy determined by the owner, allowing objects like files and printers to be created and accessed by the owner. The owner decides which users are allowed to have access to the objects and what level of access they may have. Access control lists (ACLs) store the levels of access, or permissions, which are stored in the DAC model. DAC was originally described in The Orange Book, the Trusted Computer System Evaluation Criteria, developed by the U.S. Department of Defense. The standard was superseded in 2005 by the Common Criteria for Information Technology Security Evaluation. The DAC methodology continues to be used in many personal computers and client/server networks. In Windows networks, access to objects is based on which user created them and what permissions they assign to those objects. DAC is a decentralized administration model, with administrators having control over most or all resources, while owners retain some power over their own resources.

Mandatory Access Control

Mandatory access control (MAC) is an access control policy determined by a computer system, not by a user or owner. It is used in highly classified government and military multilevel systems, but lesser implementations can be found in today's more common operating systems. MAC defines sensitivity labels assigned to subjects and objects, dictating their security level and clearance level. Access controls in MAC systems are based on data classification and "need-to-know" information, allowing users to access only what the system considers necessary. Data import and export are controlled in MAC systems. Examples of MAC include FreeBSD, Security-Enhanced Linux (SELinux), and The Orange Book. MAC is implemented in two ways: rule-based access control, which determines whether access should be granted or denied to objects, and lattice-based access control, which uses advanced mathematics to determine object access by subjects. Other related models include Bell-LaPadula, Biba, and Clark-Wilson.

Role-Based Access Control (RBAC)

Role-based access control (RBAC) is an access model controlled by the system, not by the owner of a resource. It uses sets of permissions, known as roles, to grant access to resources. Roles can be created for various job functions within an organization, with overlapping privileges and responsibilities. Administrators can create role hierarchies to define roles that can contain other roles or have exclusive attributes. RBAC is increasingly used in database access and is becoming more common in the healthcare industry and government.

Attribute-based Access Control (ABAC)

Attribute-based access control (ABAC) is a dynamic, context-aware access model that grants rights to users through multiple policies combining user, group, and resource attributes. It uses IF-THEN statements based on the user and requested resource. ABAC is becoming more flexible as technologies and organizations adopt a context-sensitive approach to authentication and access control. Access control can be centralized or decentralized, with centralized control involving one entity administering access to resources.

Access Control Wise Practices

To improve your access control plan, consider incorporating concepts such as implicit deny, least privilege, separation of duties, and job rotation. Implicit deny denies all traffic to a resource unless the users generating that traffic are specifically granted access. This is a default setting for access control lists on Cisco routers and Microsoft computers to some extent.

Least privilege is when users are given only the amount of privileges needed to do their job, reducing CPU power needed for better system stability and security. This principle is also used by programmers when developing applications and operating systems

Separation of duties is when more than one person is required to complete a task or operation. This can become a security risk if one person has too much control and completes too many portions of a task. Checks and balances are employed to maintain the proper equilibrium of users.

Job rotation is another check and balance that enforces the proper separation of duties. It creates a pool of people that can do an individual job, discourages hoarding of information, and increases operation security. By cross-training people in each department, you can defend against fraud and increase awareness, making it easier to detect if it does happen.

By incorporating these concepts, your total access control plan can be greatly improved. These principles can be applied to desktop computers and mobile devices, but the specific way they are applied will depend on the specific operating systems and organizational policies.

Rights, Permissions, and Policies

Key Topic!

Implementing a plan for access control involves strategically setting up organizational units, users, and groups, assigning permissions, and creating a safe working area to protect network data.

Users, Groups, and Permissions

User accounts can be added to individual computers or networks, such as Windows clients, Linux computers, or Macs. In larger networks with a controlling server, such as a Windows domain controller, users can access one or more computers on the domain. Users can be added in Active Directory (ADUC) or within organizational units (OUs), and user rights can be modified within the user's Properties window. Time-of-day restrictions can be set by clicking the Logon Hours button, allowing users to log on only at specific times throughout the week.

Some users have multiple accounts to access multiple systems or resources, which can lead to various issues. To mitigate these problems, consider consolidating accounts using a federated identity management (FIM) system or credential management. Security administrators should also consider using roles (RBAC) and user groups to classify users and simplify administration when assigning permissions. Groups can save time by grouping users together, allowing for easier permission assignment to files and resources.

Permissions, such as file and printer access, can be assigned to individual users or groups, and are examples of access control lists (ACLs). ACLs are lists of permissions attached to an object and reside on firewalls, routers, and computers. In Windows, ACLs are broken down into individual access control entries (ACEs), while Microsoft uses the terms discretionary access control list (DACL) and system access control list (SACL) to refer to ACLs that enable admins to log attempts to access a secured object. Sharing permissions are basic permissions applied to folders only, while NTFS permissions, also known as security permissions, can secure folders and individual files. In Linux, permissions are broken down into read, write, and execute (R, W, and X) and can be assigned to three different permission groups: owner, group, and all users. The chord command is a great way to assign permissions to all three groups using a single command. To avoid access violations, the "least privilege" or "minimal privilege" concept should be implemented, giving users only the amount of access they absolutely need.

Note!

Permission Inheritance & Propagation

Folders automatically inherit permissions from the parent folder, which ultimately comes from the root folder. This means any permissions added or removed in the parent folder will also be added or removed in the current folder. To modify these permissions, you must disable inheritance. You can copy or remove permissions from the parent to the current folder. You can also propagate permission changes to subfolders not inheriting from the current folder by selecting the Replace All Child Object Permission Entries... checkbox. In a Windows environment, there are two types of permissions: share permissions and NTFS permissions.

Quick Hits!

Moving & Copying Folders and Files

This text explains the differences between moving and copying folders in Microsoft Office. When copying a folder to the same volume or different volume, it inherits the parent folder's permissions. When moving a folder to a different location on the same volume, it retains its original permissions. Data within a volume is modified, and permissions remain the same.

Quick Hits!

Usernames and Passwords

A strong password is crucial for protecting user accounts and preventing data breaches. Passwords should contain uppercase letters, lowercase letters, numbers, special characters, and should be 8 to 10 characters or more. Some organizations require 15 characters for sensitive data.

Data exfiltration, also known as data extrusion, data exportation, or data theft, can be prevented by using complex password policies and regularly changing passwords. To protect the Administrator account, rename and password protect it by navigating to Computer Management > System Tools > Local Users and Groups > Users. Verify that the Guest account and other unnecessary accounts are disabled by right-clicking the account in question, selecting Properties, and selecting the checkbox named Account Is Disabled.

Use Ctrl+Alt+Del before the logon to add a layer of security to the logon process by ensuring users communicate through a trusted path when entering passwords. This can be added in Windows 10 by typing netplwiz or as a policy on individual Windows computers within the Local Group Policy Editor.

Use policies to enforce rules, such as self-service password management systems for large organizations with many users. These policies ensure that users reset their own passwords after a given amount of time, preventing data theft.

Policies

Policies are guidelines used to guide decisions and achieve outcomes, and can be written or configured on a computer. Local computer policies and network policies are essential for an access control model's effectiveness. Password policies can be implemented to enforce complex password usage and regulate password length. They can be configured on local computers or for an entire network, such as a Microsoft domain. The main group policy is the Default Domain Policy.

Password policies include enforcing password history, maximum and minimum password age, minimum password length, and complexity requirements. To prevent users from reusing the same password, security administrators should combine the Enforce Password History policy with the Minimum Password Age policy. Additionally, policies may require users to change passwords more than once a day to prevent bypassing the password history policy. All policies, when enabled, affect all users to which the policy applies.

Password complexity and length can be represented by equations like 26n, where 26 refers to the alphabet's 26 characters. The superscript n is a variable that refers to the length of the password, and when calculating a password, the number of characters should be raised to a particular power equal to the length of the password. For example, a password that is ten characters long would be 26 to the power of 10, or 2610. Other policies can be configured on a domain, such as the Account Lockout Threshold, which limits the number of attempts allowed when typing in a password. Kerberos protocol adds a layer of protection for username and password as they are authenticated across the network. Policies can also be configured to force locking after a certain amount of time has elapsed. Policies can be developed on all kinds of software and systems, not just operating systems. For example, organizations can implement policies on interactive bulletin board systems to prevent malicious activity.

User Account Control (UAC)

Always Remember!

User Account Control (UAC) is a Windows security feature that keeps users in standard user mode instead of full administrator rights, preventing unauthorized access and user error. Enabled, users can perform tasks as non-administrators or administrators without needing to switch users or log off.

UAC aims to eliminate unnecessary requests for excessive administrative-level access and reduce the risk of malicious software infecting operating system files. UAC can be disabled in the Control Panel, but it is not recommended. Other account prohibitions work similarly, and third-party tools are available for Windows and Linux.

CHAPTER 12:

ACRONYMS	LIST OF ABREVIATIONS
CIA Triad	*Confidentiality, Integrity, Availability*
DoD	*Department of Defense*
IA	*Information Assurance*
SLE	*Single Loss Expectancy*
ARO	*Annualized Rate of Occurrence*
DoD	*Department of Defense*
ALE	*Annualized Loss Expectancy*
MTBF	*Mean Time Between Failures*
MTTR	*Mean Time To Repair*
MTTF	*Mean Time To Failure*
FIT	*Failure In Time*
NIPS	*Network Intrusion Prevention System*
NIDS	*Network Intrusion Detection System*
NAC	*Network Access Control*
DLP	*Data Loss Prevention*
DLL	*Dynamic Link Library*
OVAL	*Open Vulnerability and Assessment Language*
OSI	*Open Systems Interconnection*
ICMP	*Internet Control Message Protocol*
HTTP	*Hypertext Transfer Protocol*
FTP	*File Transfer Protocol*
P2P	*Peer-to-Peer*
GUI	*Graphical User Interface*

Chapter 12

The Risk Management Approach to Computer Security

Security is crucial for managing the health of computers and networks. Vulnerabilities, such as open ports, weak passwords, outdated systems, and lack of policy, invite threats like malicious attacks. These vulnerabilities can also be exposed to environmental or natural threats, which are discussed in Chapters 16, "Redundancy and Disaster Recovery," and "Policies and Procedures."

Key Topic!

Computers and networks are vulnerable to various threats, including vulnerabilities. A security administrator's main task is to identify and secure these weaknesses within acceptable parameters. Vulnerabilities include open ports, unnecessary services, weak passwords, outdated systems, and lack of policy. Other threats include environmental or natural threats. Vulnerability assessment is part of overall risk management, which includes computer vulnerabilities, potential dangers, hardware and software failure, wasted time, and monetary loss.

Conducting Risk Assessments

Risk management in computer security involves identifying, assessing, and prioritizing potential threats to computer systems or networks. Smart organizations prioritize managing vulnerabilities and risk, particularly in computer hardware and software. Information assurance (IA) models, such as the CIA-(Confidential Integrity Availability) triad and DoD "Five Pillars of IA," focus on confidentiality, integrity, availability, authentication, and non-repudiation.

When handling a specific risk, organizations often choose one of the four main methodologies listed below:

- Assign the risk to a different company or individual.
- Refrain from taking a chance.
- Minimize the danger.
- Take up part of the risk, or all of it.

Risk transference involves transferring risk to a third party, such as purchasing insurance for a data center group of servers. The organization still bears the risk of losing data in case of failure.

Risk avoidance involves organizations not implementing a proposed plan due to a high risk factor, such as avoiding controversial websites due to potential attackers.

Risk management aims to reduce all risk to an acceptable level for an organization. Budgeting and IT resources dictate risk reduction and deterrents. Common methods include installing antivirus/firewall software, but high-end hardware-based solutions are less common due to cost and administration.

Risk acceptance and retention are key decisions in organizations, involving IT budgeting and resource management to manage vulnerabilities and mitigate potential threats.

After implementing risk transference, avoidance, and reduction techniques, an organization may still have residual risk. Senior management collectively decides on the residual risk in infrastructure and data, often without a single person, ensuring a comprehensive and effective risk management strategy.

Computers and computer networks are subject to a wide range of dangers. Of course, you must first evaluate the hazards in question before determining how to respond to them.

The goal of risk assessment is to estimate the potential dangers and hazards to your systems and networks that could materialize over a specific period of time. When evaluating risks, known dangers are frequently included, but new threats that may emerge might also be considered. Once the risk has been evaluated, it can be reduced until the company decides to take on more risk. Risk evaluations often go in a certain order, such as this:

- **Step 1:** List the resources owned by the company.
- **Step 2**: Determine your weaknesses.
- **Step 3:** Determine the likelihood and nature of threats.
- **Step 4:** Determine any possible financial effect.

The impact assessment is another name for the fourth stage. This is the time to calculate the possible financial expenses associated with a danger. Refer to the "Vulnerability Management" section later in this chapter for further details on Steps 2 and 3, including ways to reduce possible risks.

A risk register, or risk log, is a valuable tool for risk assessment, particularly useful for organizations with a supply chain. It helps track issues and address problems, and can be particularly beneficial for companies collaborating with suppliers and distributors to reduce risk.

The two most popular approaches for assessing risk are quantitative and qualitative. Now let's talk about these.

Qualitative Risk Assessment

In a qualitative risk assessment, the likelihood of a risk and its potential effects on the system or network are given numerical numbers. It does not place a monetary value on assets or potential losses, in contrast to its counterpart, quantitative risk assessment. It's the quicker, simpler, and less expensive method of evaluating risk, but it can't value an asset or provide an estimate of the overall amount of potential damage.

The method assigns risk ranges from 1 to 100, with higher numbers indicating higher probability of risk or greater impact on the system. For instance, a computer without antivirus software connected to the Internet has a high probability of risk and a significant impact. The qualitative risk is determined by multiplying the probability of risk and impact of the risk, resulting in 8910 out of 10,000 possible risks. Mitigation involves reducing or eliminating risks, such as installing antivirus software and ensuring auto-update.

The primary problem with this kind of risk assessment is that many different kinds of hazards are difficult to accurately quantify. The kind of qualitative system differs from one organization to the next and even across individuals; it is also a frequently discussed topic. Because of this, qualitative risk evaluations are less really measurable and more descriptive. You may, however, gain a general understanding of the danger by depending on corporate history, group polls, and own experience.

Quantitative Risk Assessment

Exact monetary values are used in quantitative risk assessment to quantify risk. It makes an effort to provide an anticipated yearly loss in US dollars for each risk. Additionally, it defines the asset values of routers, servers, and other network hardware.

When calculating quantitative risk, three values are used:

- Single loss expectancy (SLE): The monetary loss resulting from a single occurrence.
- Annualized rate of occurrence (ARO): The frequency with which a particular incidence happens annually.
- Annualized loss expectancy (ALE): The whole monetary loss incurred annually as a result of a certain incidence.

This figure represents the overall loss in dollars for that specific sort of occurrence, regardless of how many times it occurs. The following computation is used to calculate it:

$$SLE \times ARO = ALE$$

To calculate the cost of an e-commerce web server's downtime per year, we need information on the average downtime, sale amount, and sales made per minute. For example, if the server failed 7 times last year, the average downtime was 45 minutes, resulting in 315 minutes of downtime per year, close to 99.9% uptime. The average revenue per minute is $350, and a single downtime costs $15,750. To calculate the annualized loss expectation (ALE), we multiply the SLE ($15,750) by the annualized rate of occurrence (ARO).

Example of Quantitative Risk Assessment

SLE	ARO	ALE
$15,750	7	$110,250
Every web server failure results in lost revenue.	All web servers that have failed in the last year	Annual total lost as a result of a web server failure

To increase e-commerce web server uptime, organizations need to scrutinize the server's vulnerabilities and exploit them. By analyzing logs, configurations, policies, and using security tools, they can determine why the server fails frequently. This analysis is crucial as the server is crucial for the company's business.

New technologies should be carefully monitored and failures documented, using a spreadsheet to track failure history, SLE, ARO, ALE, mitigation techniques, and implementation dates, as assigning specific ALE to future incidents is not possible.

Mean time between failures (MTBF) is a reliable term that quantifies the average number of failures per million hours of operation for a product, aiding in quantitative assessments.

Note: Failure in time (FIT), which is the number of failures per billion hours of operation, is another term for maximum maintenance time (MTBF).

MTBF, mean time to repair (MTTR) and mean time to failure (MTTF) are related concepts that should be considered in disaster recovery plans, as discussed in Chapter 16.

Qualitative and quantitative risk assessment methods help predict future events, with quantitative methods being more effective. Table 12-2 provides a summary of these types.

Summary of Risk Assessment Types

Note!

Risk Assessment Type	Description	Key Points
Qualitative risk assessment	Gives numerical numbers to a risk's likelihood and potential effects on a system or network.	Numbers are arbitrary. Examples: 1–10 or 1–100.
Quantitative risk assessment	Calculates risk using precise monetary values. It makes an effort to provide an estimated yearly loss in US dollars for each risk.	Values are specific c monetary amounts. SLE × ARO = ALE MTBF can be used for additional data.

Note: The majority of businesses in the banking, pharmaceutical, and medical sectors employ quantitative risk assessments since these businesses require precise monetary values in order to quantify risk. To take things a step further, a lot of banks follow the guidelines in Basel I, II, and III agreements. The proposed guidelines delineate the appropriate capital that a bank ought to set aside to mitigate potential financial and operational risks.

Security Analysis Methodologies

To properly assess risk, analyze the security of computers, servers, and network devices. Before analysis, back up the device with files, images, or firmware. Analyze hosts to determine firewalls, configurations, anti-malware software, and updates. Develop a list of vulnerabilities and monitor for threats that could exploit these vulnerabilities, whether natural or due to user error.

There are two approaches to security analysis: active and passive.

Active security analysis involves conducting hands-on tests on a system, potentially causing a loss of productivity. It involves scanning to identify open ports or network IP addresses. Backups should be made before scanning, as intrusive scanning can be detrimental to mission-critical networks. Passive security analysis may be necessary in some cases.

Passive security analysis is a non-invasive method where tests are conducted on servers, devices, and networks without affecting them. It's useful in real-time, mission-critical networks, or computer forensics, but requires system backups. It's also known as non-intrusive or non-invasive analysis.

Fingerprinting is a security method where an attacker scans hosts to identify open ports and distinguish the operating system. Active Fingerprinting involves direct connections, potentially slowing system response. Passive Fingerprinting, on the other hand, involves random network sniffing, classifying hosts based on traffic, less common in port scanners but reducing system stress.

Security analysis can be classified as passive or active reconnaissance. Passive reconnaissance involves gaining information about a target system without engaging it, like a basic port scan. Active reconnaissance exploits vulnerabilities associated with ports, used in penetration tests.

Quick Hits!

Security Controls

Chapter 1 discussed three basic security controls: physical, technical, and administrative, with additional categorical controls as described by the NIST.

- **Management controls:** Organizational management controls focus on decision-making, risk management, procedures, policies, legal and regulatory aspects, software development life cycle, computer security life cycle, information assurance, and vulnerability management/scanning.
- **Operational controls:** Controls executed by people enhance system security, including user awareness, fault tolerance, incident handling, computer support, baseline confidence development, and environmental security, requiring technical expertise.
- **Technical controls:** Computer systems execute logical and technical controls, including authentication, access control, auditing, and cryptography. These controls include firewalls, session locks, RADIUS servers, RAID 5 arrays, and least privilege implementation.
- The preceding controls are categorical once more. You should concentrate on these defensive security controls for the Security+ exam:
- **Preventive controls:** Deterrent controls, such as biometric systems, NIPSs, and RAID 1, are implemented before an incident to prevent unauthorized access, malicious activity, and data loss.
- **Detective controls:** During an event, controls like CCTV, alarms, NIDSs, and auditing are utilized to detect malicious activity.

- **Corrective controls:** Compensating controls, such as tape backup and hot sites, are used post-event to limit damage and facilitate quick recovery, ensuring fault tolerance and disaster recovery.

Compensating controls are alternative security measures used to meet impractical or difficult security requirements. They can include network access control (NAC), data loss prevention (DLP), or additional logging and auditing. However, they do not provide the same level of security as their replaced counterparts. It is important to categorize security features in a general fashion to understand and define an organization's security controls.

Quick Hits!

Vulnerability Management

Vulnerability management involves identifying and addressing software vulnerabilities in computers and networks through analysis, testing, and regular monitoring using various security tools. There are five steps to vulnerability management:

- **Step 1. Define the desired state of security:** Organizations may have written policies defining security, which can be created by the security administrator, including access control rules, device configurations, network configurations, and network documentation.
- **Step 2. Create baselines:** Once security is established, baselines, known as vulnerability assessments, are used to evaluate the current security of computers, servers, and network devices, identifying as many vulnerabilities as possible using scanning and auditing methods.

- **Step 3. Prioritize vulnerabilities:** Prioritize vulnerabilities, such as e-commerce web servers, over single client computers without antivirus software, to create a list of items to be mitigated in order.
- **Step 4. Mitigate vulnerabilities:** Prioritize vulnerabilities and mitigate them, based on acceptable risk levels. Techniques include secure code review and system architecture and design review.
- **Step 5. Monitor the environment:** After completing mitigation, regularly monitor the environment and compare results to the original baseline. Use the new results as a post-mitigation baseline for future analyses. Be cautious of false positives and use automated monitoring tools. Continuous monitoring and confirmation validation can help reduce risk.

The five-step process for managing vulnerabilities in customers involves identifying and mitigating them. Some organizations have a defined security policy, while others may not have one. Vulnerability management involves using tools like network mappers, port scanners, ping scanners, protocol analyzers, and password crackers to identify and mitigate vulnerabilities. These assessments can reveal sensitive data, open ports, weak passwords, and system failures.

Penetration Testing

Penetration testing is an active method of assessing system security by simulating attacks. It differs from regular vulnerability scanning, which may be passive or active, as it aims to determine the impact of a specific threat on an organization.

Penetrating tests can be done blindly, white-box, or gray-box, showcasing the potential for attacks. They are typically conducted on publicly accessible servers or network devices. One common technique is the pivot, which allows attackers to move to other systems after gaining access to a system. This can be prevented through host-based intrusion detection systems (IDS), secure coding, and network-based solutions.

Persistence is an exploit where an attacker attempts to reconnect at a later date using backdoors, privilege escalation, and cryptographic keys. Non-persistent systems can be developed using master images, snapshots, reverting to known states, rolling back to known good configurations, and live boot media. Race conditions, also known as time-of-check or time-of-use attacks, are difficult to perform due to the small window between service usage and security control execution. Preventing these exploits requires proper secure coding and planning of system and network architecture.

Here are a few approaches to carrying out penetration testing:

- **The Open Source Security Testing Methodology Manual (OSSTMM):** The manual and methodology outline the appropriate security testing method, adhering to the scientific method, and can be obtained from ISECOM.
- **NIST penetration testing:** SP800-115, a less detailed document from the U.S. government, is favored by many organizations due to its consistency with the OSSTMM.

Note: When combined with DLL injection testing, penetration testing may become even more invasive (active). This is the situation when dynamic link libraries are compelled to operate in the memory area that is now being utilized, altering program behavior in a way that the developer did not want or predict.

OVAL

The Open Vulnerability and Assessment Language (OVAL) is an international standard funded by the U.S. Department of Homeland Security, regulating secure public information transfer across networks and the Internet. It consists of the OVAL Language and Interpreter.

OVAL Language: OVAL is based on three distinct XML schemas that have been designed as its framework:

1. System testing information
2. System state analysis
3. Assessment results reporting

OVAL is an XML schema that defines and explains the XML documents that must be written for usage with OVAL, as opposed to a language like C++.

OVAL Interpreter: A reference designed to check against OVAL schemas and definitions to make sure the right syntax is being used. The OVAL Interpreter is linked to a number of downloads, as well as support files and discussion boards that let security professionals double-check their work for correctness.

OVAL is a tool for standardizing security advisory distributions, enabling users to regulate their advisories, conduct vulnerability assessments, manage patches, audits, and identify threats.

OVAL is utilized by Hewlett-Packard, Red Hat Inc., CA Inc., and the U.S. Army CERDEC (Communications-Electronics Research, Development and Engineering Center).

Additional Vulnerabilities

Table 12-3 outlines general vulnerabilities and prevention methods, with security tools discussed in the following section for assessing and preventing exploits.

Common Vulnerabilities and Fundamental Preventive Techniques

Vulnerability	Prevention Methods
• **Improper input handling** • **Improper error handling** • **Memory vulnerabilities**	Secure coding, SDLC (see Chapter 5)
• **Default configuration** • **Design weaknesses** • **Resource exhaustion** • **Improperly configured accounts**	Harden systems (see Chapter 4) Proper network design (see Chapter 6) Properly configure and audit permissions (see Chapter 11)
• **System sprawl** • **End-of-life systems** • **Vulnerable business processes**	Proper network auditing (see Chapter 6 and Chapter 13) SDLC (see Chapter 5) Implement secure policies (Chapter 18)
• **Weak ciphers** • **Improper certificates**	Upgrade encryption (see Chapter 14 and Chapter 9) Review certificates, use a CRL (see Chapter 15)
• **New threats** • **Zero-day attacks**	Keep abreast of latest CVEs and CWEs (see Chapter 6 and Chapter 5) Plan for unknowns (see Chapter 7)

• **Untrained users**	Educate users about social engineering methods (see Chapter 17) Educate users about malware and attacks (see Chapter 2 and Chapter 7)

Assessing Vulnerability with Security Tools

This section discusses the importance of security tools in testing, analyzing, and assessing computers and networks. Vulnerabilities can arise from backdoors, open ports, or poor design. Security administrators should have thorough computer and network documentation, using tools like Microsoft Visio and network mapping. Vulnerability scanners, protocol analyzers, and password crackers are used to assess network vulnerability. These tools are used by security administrators and hackers, with the former focusing on identifying vulnerabilities and mitigating risk, and the latter exploiting them.

Network Mapping

Network documentation is crucial for determining security levels. Network mapping software, like SolarWinds' Network Topology Mapper, helps in analyzing network connectivity. It maps layers 1 through 3 of the OSI model, providing a comprehensive view. This software is suitable for off-hours or when the network is at its lowest usage. Other free and paid network mapping software is available, and users can export mapped contents to Microsoft Visio.

Wireless networks can be surveyed using applications like Air-Magnet and NetStumbler to map wireless clients and locate available Wireless Access Points (WAPs). Manual inspection of certain areas is necessary for network documentation. Network diagramming programs like Microsoft Visio can create diagrams and flowcharts that are useful planning tools for network administrators and security people. An example of a network diagram is shown in Figure 12-1, which was created by mapping a now-defunct network, exporting results to Visio, and making manual adjustments. This documentation helps mitigate risks and serves as a foundation for security analysis.

Avoid memorizing passwords and store them on paper in a secure, non-removable safe. Keep passwords on paper and limit access to admins. Keep a list of IP addresses and computer names in Excel or Access. Network mapping can aid in vulnerability assessment, so use programs and thoroughly document your network for better analysis and baseline.

Vulnerability Scanning

Always Remember!

Vulnerability scanning is a technique that identifies threats on a network without exploiting them. To assess network vulnerability, use a general vulnerability scanner and a port scanner to determine the attack surface of systems. Nessus is a commonly used vulnerability, compliance, and compliance tool, but it can be resource-intensive. Nmap is a good port scanner for port scanning, providing a fingerprint of the operating system. It shows inbound ports and services running on a remote computer, allowing for more information about the operating system and applications. This is an example of OS fingerprinting.

Examine open ports and identify unnecessary services or processes. Disable unnecessary ones, like stopping DNS services, to prevent server vulnerabilities. Re-scan the computer after ensuring risk mitigation.

Network administrators should scan for non-essential services and close any corresponding ports to ensure security. Netcat and Ncat are useful tools for port scanning, listening, and transferring files. The netstat command displays ports in use for local and remote computers, but only for local computers. These tools help prevent malicious or malicious use of services.

Note: You may also use other simple port scanners, like Angry IP Scanner (and many more free port scanners available online). Certain tools have the ability to function as ping scanners, generating ICMP echoes in order to locate IP addresses inside a specific network segment.

Network enumerators, like Nmap and Nessus, extract information from servers, including network shares, services, and user groups. Banner grabbing, a technique used to find information about web servers, FTP servers, and mail servers, is used by network administrators to inventory systems and services. It can also be used by attackers to gather HTTP headers. It is essential for security administrators to prevent the use of banner-grabbing tools.

Network Sniffing

Protocol analyzers, packet sniffers, and network sniffers are all related terms. They are tools used to find and investigate other computers on a network, capturing packets for analysis. They provide more detailed information about traffic coming and going to a host than vulnerability scanners or port scanners.

Wireshark is a protocol analyzer that captures packets on the local computer. In a packet capture, an ICMP packet is captured from a ping request sent from a local computer to a remote host. The device, a router, may not reply to pings, potentially exposing a vulnerability. To mitigate this risk, turning off ICMP echo replies on the router is recommended.

This program can be used to identify unauthorized FTP, gaming, and P2P traffic, which consumes bandwidth and resources, and can be vulnerabilities. Removing these servers reduces risk and allows organizations to focus on work. Malicious users can use protocol analyzers to capture passwords and other confidential information. Wireshark is the industry standard for software-based packet sniffers. Hardware-based devices like Fluke Networks can analyze networks and hosts, with built-in GUI systems for monitoring ports, troubleshooting authentication issues, and identifying network resources and IP addresses.

Password Analysis

The book discusses the importance of passwords in operating systems and networks, highlighting the prevalence of weak passwords. Password-cracking programs like Cain & Abel can help crack passwords on local systems and remote devices. The program can detect weak passwords and help users forget their passwords. It can also detect if a policy is complex or outdated. To prevent this, it is recommended to scan computers and devices for weak passwords using a password cracker.

Chapter 14 discusses encryption and hashing concepts, with Cain & Abel being a free download. Other tools like ophcrack, John the Ripper, THC-Hydra, Aircrack-ng, and RainbowCrack are password/passphrase-cracking tools.

The many techniques for breaking passwords are listed in the following list. Password recovery, often known as cracking, can be accomplished in many ways:

- **Guessing:** Weak passwords can be guessed by smart people, especially if they know the user. Common passwords include password, admin, secret, and more. An attacker could use the user's username, date of birth, or reverse letters. Although not a technical method, guessing exposes many passwords daily worldwide.
- **Dictionary attack:** Tries each word on a predetermined list of potential words one at a time.
- Keys, passphrases, and passwords can all be cracked using it. It functions best when attacking numerous systems and using weak passwords. The strength of the vocabulary that the password-cracking software uses determines how powerful the dictionary assault will be.
- **Brute-force attack:** Password-cracking is a last resort method that attempts every possible password instance, working best on shorter passwords but potentially breaking any password with enough time and CPU power. It depends on whether the attack is online or offline, with online attacks having a short window and offline attacks having more time. Some programs use dictionary and brute-force attacks.

Note: A program that combines brute-force and dictionary assaults in a hybrid attack is what some attackers will use.

- **Cryptanalysis attack:** Rainbow tables, or precalculated encrypted passwords, are used in password attacks like precomputation. These attacks hash and store words in a dictionary to recover passwords quicker. Salting, randomization of the hashing process, can defeat this attack.

Knowledgeable attackers understand where password information is saved. In Windows, it is stored in an encrypted binary format within the SAM hive. In Linux, the data used to verify passwords was historically stored in the

/etc/passwd file, but in newer Linux systems the passwd fi le only shows an X, and the real password information is stored in another file, perhaps /etc/shadow, or elsewhere in an encrypted format.

Passwords can be obtained through various methods, including viruses, Trojans, wiretapping, and phishing. Protecting passwords is crucial, as it's just as important as creating complex password policies and periodically monitoring and updating them.

For the Security+ exam, a basic understanding of password cracking is sufficient, as there are numerous types of hashes, algorithms, and tools available.

CHAPTER 13:

ACRONYMS	LIST OF ABBREVIATIONS
SNMP	*Simple Network Management Protocol*
IDS	*Intrusion Detection System*
IPS	*Intrusion Prevention System*
RTOS	*Real-Time Operating System*
TLS	*Transport Layer Security*
SNMP	*Simple Network Management Protocol*
IDS	*Intrusion Detection System*
SSL	*Secure Sockets Layer*
SYN	*Synchronize*
ACK	*Acknowledgement*
FTP	*File Transfer Protocol*
TCP	*Transmission Control Protocol*
UDP	*User Datagram Protocol*
SQL	*Structured Query Language*
DNS	*Domain Name System*
Syslog	*System Logging Protocol*
WORM	*Write Once Read Many*
SIEM	*Security Information and Event Management*

Chapter 13

Monitoring & Auditing

Note!

This chapter discusses monitoring and auditing, focusing on methodologies and tools for computer security audits. It assumes understanding Chapter 12, "Vulnerability and Risk Assessment," and employing the concepts and tools. The chapter emphasizes the importance of putting on a sleuthing hat to solve networking and operating system security issues, using the adage "When you have eliminated the impossible, whatever remains, however, must be the truth."

Monitoring Methodologies

To maintain a secure network, it's essential to monitor all systems, applications, servers, and devices. This surveillance increases the security of the entire infrastructure and helps analyze legitimate events. Monitoring can be done manually or automated, with manual methods involving analyzing log files and policies. Automated monitoring, such as data mining programs and antivirus, can scan for errors, malicious attacks, and anomalies. The three main types of automated monitoring are signature-based, anomaly-based, and behavior-based.

Signature-Based Monitoring

Signature-based monitoring analyzes network traffic for predetermined attack patterns, stored in a database. These signatures must be updated regularly to ensure network security. However, only specific attacks match the signature, making it vulnerable to false negatives. To protect against this, the system should be updated with the latest signatures. Some signature-based monitoring systems use heuristic signatures, which are CPU-intensive and require fine-tuning for specific networking environments.

Anomaly-Based Monitoring

An anomaly-based monitoring system establishes a performance baseline based on normal network traffic evaluations during regular working hours. It compares current traffic to the baseline to detect if it's within the parameters. If it's outside, an alarm is triggered. The accuracy of the baseline is crucial, as inaccurate baselines increase the likelihood of false indicators, such as false positives, which can be detected through improperly configured IDS or IPS solutions.

Behavior-Based Monitoring

Behavior-based monitoring systems compare previous behavior of applications, executables, and the operating system to current activity, attempting to stop improper behavior if necessary. This approach has advantages over signature-based and anomaly-based monitoring but can lead to false positives due to numerous application types and relationships. Heuristic analysis, often used in antivirus software and IDS/IPS solutions, detects malicious behavior without uniquely identifying it, but can lead to low accuracy and high false positives.

Using Tools to Monitor Systems and Networks

Monitoring tools and creating baselines are crucial for network security. Performance monitoring software, protocol analyzers, and analytical utilities can help quickly detect and mitigate threats. Common tools include Windows' Performance tool, Wireshark protocol analyzer, and other command-line and GUI analytical tools.

Performance Baselining

Baselining is the process of measuring changes in networking, hardware, software, and applications to identify the security posture of an application, system, or network. Baseline reporting helps security administrators identify vulnerabilities and weaknesses in systems. A security baseline defines the basic set of security objectives that must be met by any given service or system. Baseline deviations should be investigated immediately, especially for mission-critical systems and RTOS-based computers.

Creating a baseline involves selecting something to measure and consistently measuring it for a period of time. For example, measuring the average hourly data transfer to and from a server's network interface can be done using a performance monitoring tool or protocol analyzer. By averaging the results together, a baseline is created, which can be compared to future measurements to determine the standard load of the server and the requirements needed on a consistent basis.

Baselining can be used for performance monitoring and benchmarking, such as measuring network activity, CPU usage, memory, hard drive resources used, and more. Tools like Performance Monitor in Windows can help create baselines by measuring network activity, CPU usage, memory, and hard drive resources. However, it is important to measure the same way at the same time each day or week to provide accurate comparisons.

Setting up alerts is also wise, as they can be generated automatically by the system and sent to administrators and other IT people. Alerts can be set off in various ways, such as if the CPU exceeds a certain threshold or if the physical disk is peaking at 100 MB/s for more than 5 minutes.

Protocol Analyzers

Protocol analyzers are essential tools for network security, capturing and analyzing packet traffic. They can detect broadcast storms, which are accumulated packet traffic from network interfaces, and identify header manipulation, which can lead to XSS attacks, poisoning attacks, hijacking, and cookie manipulation. Protocol analyzers can also analyze TCP/IP handshakes, revealing attacks like TCP hijacking. They can detect synchronized sequence numbers (SYN) and acknowledgement field significant (ACK), which can be exploited by attackers. Wireshark is an industry standard protocol analyzer that can help identify and prevent these attacks. Overall, protocol analyzers are essential for network security.

Wireshark

Quick Hits!

Wireshark is a free software that captures packets on the local computer installed on it, allowing it to detect vulnerabilities and monitor both local and remote systems. It works in promiscuous mode, allowing it to delve into packets even if they weren't addressed to the computer it runs on. To monitor remote systems, start sessions from the client computer to those systems and monitor the packet stream. If installing Wireshark on a server, you will need the latest version of WinPcap.

To use Wireshark, you can use it on a client computer, initiate a packet capture, and verify that TLS 1.0 is being used. If you notice other protocols like SSL 2.0 that raise red flags, investigate further, leading to protocol upgrades or changes.

Wideshark can be used with a network adapter configured for promiscuous mode. Port mirroring or a network tap can be used to capture and analyze packets, but they can place a significant load on the CPU and can lead to dropped packets.

To monitor an FTP server, use a packet sniffer to capture packets on the monitoring server and log in to the server. Increase security on the FTP server and use more secure programs, such as domain-based authentication or FileZilla or WS_FTP.

SNMP

The Simple Network Management Protocol (SNMP) is a TCP/IP protocol used to monitor network-attached devices and computers. It is typically integrated into a network management system or free monitoring software. SNMP consists of three components: managed devices, agents, and the network management system. SNMP uses ports 161 and 162 to send requests and notifications. Applications using SNMP versions 1 and 2 are less secure and should be replaced with software supporting SNMP version 3. SNMPv3 provides security through encrypted packets and additional message integrity.

However, there are vulnerabilities, such as weak symmetric algorithms in power distribution units. Historically, SNMP-based monitoring tools have been in-band, but out-of-band management is necessary for devices without direct network connections. In mission-critical networks, out-of-band management tools are crucial, as they provide an independent access switch for managed devices. The decision should be carefully considered, considering the security risks associated with each additional system or network.

Analytical Tools

This book discusses the use of various analytical tools for security purposes, including monitoring, auditing, vulnerability assessment, and forensics. Open sessions and files can be monitored within Computer Management in Windows, where you can see shared files and network sessions. The Performance Monitor utility can also be found within Computer Management.

The openfiles command can be used to view files opened by remote computers and disconnect them. The FC command can be used to compare files suspected of compromise with older versions. Files can also be viewed and closed with the net file command.

Net config, net session, and net view commands can be used to monitor files. The netstat command can show open ports in numeric format and the amount of bytes and packets sent and received. Linux commands can view processes, services, and open files.

Keyloggers, both hardware and software-based, can be used to monitor individual keystrokes stored on servers. Hardware devices are inline devices that connect to the end of a keyboard's cable and can store gigabytes of data and transmit data wirelessly. Keyloggers can be identified by their identical ports on each end and can be detected using software tools.

Anti-keylogger software and security tokens can be used to prevent these tools, which can be detected using live USB/CD/DVD operating systems and network monitoring programs.

Use Static and Dynamic Tools

Analytical tools, both integrated into the operating system and offered by third parties, are static and not suitable for monitoring environments. Dynamic tools like Performance Monitor and Wireshark are better. However, snapshot tools like openfiles provide a quick overview. Combining static and dynamic tools is essential for auditing.

Quick Hits!

Conducting Audits

Computer security audits are technical assessments conducted on applications, systems, or networks, focusing on detective security control. They can be done manually or with computer programs, including reviewing security logs, access control lists, user rights, permissions, group policies, vulnerability scans, written organization policies, and interviewing personnel. Auditing files can be broken down into a three-step process: turning on an auditing policy, enabling auditing for specific objects, and reviewing the security logs to determine who did what to a resource and when.

A security administrator is most concerned with auditing files, logs, and systems security settings. They need to determine who did what to a resource and when, and whether users have succeeded or failed in their attempts to access, modify, or delete objects. Security logs act as proof that a user account was involved, ensuring non-repudiation.

A common problem with security logs is that they fail to become populated, especially on older systems. To address this, it is essential to check if auditing has been turned on in the correct policy, if the person attempting to view the log has administrative capabilities, and if the audit recording has failed.

Logging

Logging is a crucial method for auditing organized information, and frequent monitoring of logs is essential for security professionals. The Security log in Windows is the most important log file, as it shows whether a user was successful at various tasks such as logging on to the local computer or domain, accessing, modifying, or deleting files, and modifying policies. These events are called audit trails and should be monitored frequently to keep on top of any security breaches.

Other types of Windows log files should be monitored periodically, including System logs, Application logs, and DNS logs. System and Application logs exist on client and server versions of Windows, while Server logs exist only on servers. Examining DNS logs can uncover unauthorized zone transfers and other malicious or inadvertent activity on the DNS server. Web servers can also be analyzed and monitored to determine if the server has been compromised.

Other operating systems, applications, and devices have their own set of log files, such as Microsoft Exchange and SQL database servers, and firewalls. The firewall log is especially important, as it can show malicious port scans and other vulnerability scans. Syslog is a useful tool for security administrators, as it is the standard for computer message logging and can be exported in real time to a computer running a Syslog server.

The Syslog server can run directly on your workstation and pull logs from devices, allowing you to easily monitor what is happening on those devices from the comfort of your seat. By default, Syslog uses port 514 and works over a UDP transport mechanism. Some companies offer Syslog programs, such as SolarWinds Kiwi and Syslog Watcher, which work in basically the same way but may use a proprietary port number instead of the standard 514 and may offer TCP connectivity to avoid packet loss.

Utilities can be downloaded that convert event logs from the Event Viewer into Syslog messages.

Log File Maintenance and Security

A must know!

Log files should be carefully planned, maintained, and secured. Factors to consider include storage, backup, and securing. Before setting up a logging system, consider the amount of disk space needed, the information needed to reconstruct logged events, and the level of detail allowed. Verbose logging is used to gather as much information as possible, and time stamping is incorporated to reduce CPU usage. Log files can be saved to different partitions or systems, depending on the system's speed. The size and overwriting configuration of the file should also be considered. For security, logs should not be overwritten, and administrators should save and back up the log monthly or weekly. If the log becomes full, an alert should be set up to notify the administrator. Backups should be done to a remote log server, separate physical offsite locations, or WORM media types. Retention policies should be in place for log files. Securing log files can be done through backup methods and setting permissions to the file. File integrity is also crucial, and encryption through hashing can verify the integrity of the log files.

Auditing System Security Settings

Auditing system security settings is crucial, particularly for user permissions and group policies. Shared folders on the network should be reviewed, as they should be located on a different partition, drive, or computer. Some companies opt to secure administrative shares, such as IPC$ and ADMIN$, for client computers. Verify proper ownership of the folder and ensure no one else has inadvertently or maliciously taken control. In the Advanced Security Settings dialog box, check if auditing settings correspond to an organization's written policies.

Computer policies should also be reviewed, as there might be different policies for each department in an organization. If a security template is used, verify that it is being used and that the settings include what the organization has defined as part of its security plan. If a template is not used, go through as many policy objects as possible, including password policy, security options, and audit policy. Spot checks should be made of individual computers to ensure they adhere to the rules.

SIEM

Always Remember!

SIEM solutions, like HPE's ArcSight and IBM's QRadar, offer real-time monitoring of systems and logs, automation, and data aggregation. They also provide correlation engines, compliance with government auditing processes, and forensic analysis. SIEM also includes WORM functionality, automatic deduplication, and scanning for confidentiality compliance, ensuring information cannot be modified.

CHAPTER 14:

ACRONYMS	LIST OF ABBREVIATIONS
AES	*Advanced Encryption Standard*
DES	*Data Encryption Standard*
3DES	*Triple DES*
RC	*Rivest Cipher (also humorously referred to as "Ron's Code")*
IDEA	*International Data Encryption Algorithm*
RSA	*Rivest-Shamir-Adleman*
TKIP	*Temporal Key Integrity Protocol*
WEP	*Wired Equivalent Privacy*
WPA2	*Wi-Fi Protected Access 2*
SSL	*Secure Sockets Layer*
TLS	*Transport Layer Security*
PGP	*Pretty Good Privacy*
PKI	*Public Key Infrastructure*
DEK	*Data Encryption Key*
KEK	*Key Encryption Key*
MEK	*Master Encrypting Key*
ECC	*Elliptic Curve Cryptography*
ECDH	*Elliptic Curve Diffie-Hellman*
PFS	*PFS: Perfect Forward Secrecy*
DHE	*DHE: Diffie-Hellman Ephemeral*
EDH	*EDH: Ephemeral Diffie-Hellman*
HSMs	*HSMs: Hardware Security Modules*
VoIP	*Voice over Internet Protocol*
IPSec	*IPSec: Internet Protocol Security*
QKD	*Quantum Key Distribution*
PGP	*Pretty Good Privacy*
SHA	*SHA*
MD5	*Message-Digest algorithm 5*
RIPEMD	*Race Integrity Primitives Evaluation Message Digest*
HMAC	*Hash-based Message Authentication Code*
NTLM	*Network Trust Level Manager*
LPP	*Least Privilege Principle*
LSASS	*Local Security Authority Subsystem Service*
SSH	*Secure Shell*
OTP	*One-Time Passwords*
HOTP	*HMAC-based OTP*
TKIP	*Temporal Key Integrity Protocol*
WEP	*Wired Equivalent Privacy*

Chapter 14

The Keys to Data Protection

A must Know!

Throughout history, the need to secure sensitive information has driven the development of ingenious methods. Encryption plays a crucial role in safeguarding data by transforming it into an unreadable format for unauthorized individuals. In our digital age, encryption stands as the invisible vault, safeguarding our most valuable asset: data. This chapter is designed to provide a foundational understanding of encryption concepts, methods, and types, which are fundamental for the CompTIA Security+ exam.

The importance of encryption lies in its ability to protect data throughout its digital lifecycle. We encounter data in three primary forms: data in use, data at rest, and data in transit. Data in use refers to actively processed data, while data at rest pertains to stored or archived data. Data in transit encompasses information moving across networks or residing temporarily in computer memory.

Cryptography Concepts

The digital age revolves around information, and safeguarding it is paramount. This is where cryptography steps in. Derived from the Greek words "Kryptos" (hidden) and "graphein" (writing), cryptography is the art and science of securing communication by transforming messages into an unreadable form. Unlike the simple act of hiding a phone number with a code, cryptography utilizes complex mathematical algorithms and techniques to create a robust shield for sensitive information.

At its core, cryptography functions as follows:

Imagine a person with a black book who modifies a real phone number by shifting each digit backward or forward by a specified number of steps, resulting in a coded phone number, such as 222-7053, similar to modern data encryption methods.

Black Book Phone Number Encryption

Original Number	**Modifier**	**Modified Number**
0	Minus 3	7
1		8
2		9
3		0
4		1
5		2
6		3
7		4
8		5
9		6

Every number in this example, ranging from 0 to 9, has a corresponding number three digits behind it. This "modifier" essentially tells you how many positions to shift the original digit. For example, a modifier of "-3" means you move the original digit three steps backward (around the bend if necessary) to get the encrypted version. In this case, a number like "0" with a "-3" modifier would become "7" (0 -> 9 -> 8 -> 7).

NOTE: This is based on the Caesar Cipher, or more precisely the Caesar Shift Cipher, in which letters were shifted to one or more positions in communications conveyed throughout the Roman Empire. Another cipher you may be familiar with is the ROT13 substitution cipher, which rotates or substitutes a letter with the letter that comes 13 steps after it.

Now let's make an analogy. Every element in the table is comparable to current computer-based encryption ideas:

- The modified number is like encrypted data.
- The original number is like the original file data.
- The modifier is like an encryption key.

Although I refer to this as the "Black Book Example," I'm sure that other people have made comparable comparisons. This is a very rudimentary example, of course, but it should help you make the connection between this more concrete concept and real computer-based encryption solutions.

Consequently, in order for other individuals to locate the original phone numbers in the black book, they would need to perform the following tasks:

- **Step 1**: Acquire entry to the black book. It's similar to getting access to data. This could be challenging in and of itself, depending on how securely the black book is fastened.
- **Step 2:** Decrypting data involves breaking the code. If the owner of a black book memorizes the code and keeps it secret, it becomes harder for others

to crack. However, a more advanced code, like Table 14-2, can be more challenging to crack.

Advanced Black Book Phone Number Encryption:

Original Number	Modifier	Modified Number
0	Minus 9	1
1	Minus 8	3
2	Minus 7	5
3	Minus 6	7
4	Minus 5	9
5	Minus 4	1b
6	Minus 3	3b
7	Minus 2	5b
8	Minus 1	7b
9	Minus 0	9b

This example involves a modified key for each original number, which is harder to decrypt due to its increased variations. This is harder for computers and hackers to process. A secret key is used to keep phone numbers secret from others, allowing only one person to access the encryption codes.

Key Topic!

The Building Blocks of Encryption

This section delves into the technical terms cryptography, encryption, ciphers, and keys, highlighting their importance and the need to read through them twice.

- **Cryptography**

Cryptography is the study of hiding information or meaning, commonly used in e-commerce and passwords, with encryption being the most common method for encrypting messages.

- **Encryption**

Encryption, a core tool within cryptography, focuses on transforming plain text data – your emails, messages, or files – into a scrambled mess called ciphertext. This ciphertext appears unintelligible without the proper key.

- **Cipher**

A cipher is an algorithm that performs encryption or decryption, such as stepping back three letters in the alphabet to create a ciphertext. Historical ciphers used substitution and transposition methods, but modern algorithms are more complex. Algorithms are well-defined instructions that describe computations from their initial state to their final state, and the entire set of instructions is the cipher.

- **Key**

The key is the crucial element that determines the output of a cipher, and it is essential for cipher computation. It is kept secret to provide security, and the length of the key determines its strength. Shorter, weaker keys are more vulnerable to attackers. The key exchange process in encrypted messages starts with a key exchange, which varies depending on the cryptographic algorithm.

Keys can be private or public, with private keys being secretly known to specific users, and public keys being known to all parties involved in encrypted transactions. Examples include smart cards and USB flash drives. Out-of-band key exchange involves private keys stored on devices, while in-band key exchange involves public keys for secure internet communication.

Symmetric vs. Asymmetric Key Algorithms

Cryptographic systems utilize symmetric or asymmetric keys, and understanding their differences and their combined use is crucial for effective security.

Key Topic!

Symmetric Key Algorithms

Symmetric key algorithms, also known as secret, private, single, or shared keys, are used for encryption and decryption in ciphers like DES, 3DES, RC, and AES. Kerberos, a technology that uses symmetric keys, uses a key distribution center for secure transmission.

NOTE: By using asymmetric keys, Kerberos has the option to employ public key cryptography, which is discussed later in this chapter. This is carried out at certain authentication phases. The chapter "Physical Security and Authentication Models," chapter 10, goes into further detail about Kerberos.

The private key is a common workplace security measure, used to encrypt data. If a user's account is deleted, the data can only be recovered using a recovery agent. It's crucial to approach private keys and symmetric key systems cautiously to avoid data loss.

Two categories of symmetric key algorithms are as follows:

- **Stream Cipher** encrypts each binary digit in the data stream, one bit at a time.

- **Block Cipher** encodes a collection of bits as separate units called blocks. The Advanced Encryption Standard (AES) algorithm, for instance, is capable of utilizing 256- or 128-bit block ciphers. Block ciphers can function in a variety of ways, such as: The modes of encryption include Electronic Codebook (ECB), Cipher Block Chaining (CBC), Cipher Feedback (CFB), Output Feedback (OFB), Galois/Counter Mode (GCM), and Counter (CTR). ECB divides a message into blocks of plaintext, while CBC builds on it by XORing each block with the previous ciphertext block. Secure coding concepts should be employed, or a separate block mode like GCM is recommended.

Symmetric key algorithms require a secure initial exchange of secret keys between the sender and receiver. Encrypting the exchange can be a solution, but it can be challenging depending on the circumstances. People may share their cipher with others. Symmetric encryption is preferred for large data encryption and transmission due to its faster encrypting and decryption time compared to asymmetric encryption.

Asymmetric Key Algorithms

Asymmetric key algorithms use a pair of different keys to encrypt and decrypt data, which are mathematically related but not identical. This is analogous to physical and electronic locks, where one key locks the door and the other unlocks it. Asymmetric key management schemes, like PKI, are complex due to the difference in keys. Examples include RSA, Diff-Hellman system, and elliptic curve cryptography. SSL and TLS protocols also use asymmetric key algorithms.

Public Key Cryptography

Key Topic!

Public key cryptography uses asymmetric keys, creating a secret private key and a published public key. The public key is widely known, but only the private key owner can decrypt messages. The security of the system relies on the private key's secrecy.

Public key cryptography uses a private key to sign a message, which can be checked with the public key using a digital signature. This ensures integrity, non-repudiation, and protection against forgery and tampering. Asymmetric keys function encrypt, sign, decrypt, and verify.

NOTE: In order to compare digital signatures after the document reaches its final destination, they can additionally be hashed (more on hashing later).

The Diffi e-Hellman scheme involves users generating public/private key pairs and sharing them with others. This creates a shared secret for a symmetric cipher. Public key cryptography can use asymmetric keys alone or in combination.

Internet standards like SSL/TLS and PGP use public key cryptography, not public key infrastructure (PKI). PKI is a system of hardware, software, and policies that binds public keys with user identities through certificates and a certificate authority.

Key Topic!

Key Management

Key management is crucial for maintaining secure relationships between users and keys, including generation, exchange, storage, and usage. It involves choosing and securely storing keys, with strong keys being the most important concept. Exchange methods include encapsulating keys, using key indicators, and exchanging symmetric session keys. Proper storage depends on users and passwords, and keys should be replaced frequently to prevent cracking.

Steganography

Key Topic!

Steganography is a form of security through obscurity, where messages are written on a person's scalp with indelible ink. The goal is to make no one suspect the existence of the hidden message, allowing the sender and receiver to remain unaware. Steganography is often used in conjunction with cryptography, as seen in ancient Greece, where a messenger would cover their head to hide the message.

Steganography conceals messages within encrypted documents, sound files, images, slowed-down videos, Word documents, Excel spreadsheets, VoIP conversations (Lost Audio Packets Steganography), streaming services, and compromised wireless networks using the HICCUPS system.

Steganography is a technique used to send hidden messages through graphic files. It involves replacing the least significant bit of each byte, such as shading the color of a pixel or triad, to alter the binary number associated with the color. This process requires the recipient to have the original file and can be automated using programs on the Internet. Steganography aims to provide obfuscation, making something unclear and difficult to undo manually.

Encryption Algorithms

Now that we've established the core concepts of cryptography, encryption, ciphers, and keys, it's time to delve deeper into the fascinating world of encryption algorithms. These algorithms are the secret recipes, the complex mathematical formulas that transform our plain text data into unreadable ciphertext and back again.

DES and 3DES

The Data Encryption Standard (DES) was the U.S. government's encryption standard in the 1970s, but due to its weak key and 56-bit key size, it is now considered deprecated. It was replaced by Triple DES (3DES) in 1999, and the Data Encryption Algorithm (DEA) is based on the Feistel cipher, reducing code requirements.

NOTE: Although it is an optional method in the OpenPGP standard, the International Data Encryption method (IDEA) was created as DES's replacement. However, its basic key schedule produces weak keys.

Triple DES, also known as 3DES or TDEA, applies a cipher algorithm three times to each cipher block, allowing a key size up to 168-bit, but has been overshadowed by AES since late 2001.

AES

In the late 1990s, the National Institute of Standards and Technology (NIST) launched a competition for advanced encryption, with Rijndael winning. This submission was later developed into the Advanced Encryption Standard (AES), which became the U.S. government standard in 2002.

AES is a fast, efficient encryption algorithm that uses a substitution-permutation network to create ciphertext. It uses 10, 12, or 14 rounds for each version. AES is suitable for various platforms, including wireless networks running the WPA2 protocol, remote control applications, and memory encryption. The IEEE 802.11i standard specifies AES usage with WPA2, deprecating WEP and WPA. AES is also used in memory encryption for passwords and other PII.

AES is a strong type of symmetric encryption, suitable for transferring encrypted data to USB flash drives and whole disk encryption techniques like BitLocker. It is susceptible to related-key attacks and side-channel attacks, but remains widely used and uncompromised, with some experts believing it will never be compromised.

DEK, KEK, and MEK are keys used in encryption, specifically AES. A data encryption key (DEK) is used to encrypt data, while a key encryption key (KEK) stores the DEK in an encrypted format. A master encrypting key (MEK) is another type that describes either a DEK or KEK. In secure storage scenarios, a DEK is used to encrypt data in a user's protected area, encrypted by a KEK based on the user's password.

RC

The abbreviation "RC" can be a bit confusing in the world of encryption. It officially stands for "Rivest Cipher," named after its creator Ron Rivest. But you might also hear it jokingly referred to as "Ron's Code." There are actually several different versions of RC out there, and most of them aren't related to each other beyond the fact that they're all encryption methods.

One specific version, RC4, has been widely used in the past for protocols like SSL (used for secure websites), WEP (wireless network security), and RDP (remote desktop access). However, RC4 has some known weaknesses, which means it's no longer considered secure. There are temporary workarounds like TKIP, but the current recommendations are to use stronger encryption methods like AES and WPA2 for these purposes. Even Microsoft's Remote Desktop

Services uses RC4 by default, but it's best to disable it and switch to a more secure, government-approved (FIPS-compliant) encryption option.

NOTE: A crypto-module that is FIPS-compliant may be software- or hardware-based. Microsoft is listed as a proper vendor, for instance, in FIPS 140-1 and 140-2; for additional information, visit http://csrc.nist.gov/groups/STM/cmvp/documents/140-1/140val-all.htm. Microsoft offers a system program interface (SPI) called Microsoft Cryptographic Service Provider (CSP), which is implemented by a software library.

RC5 is a block cipher known for its simplicity and variable size. The strongest RC5 block cipher was cracked by distributed.net in 2001 with a 64-bit key. RC6 is a patented algorithm offered by RSA Security as an alternative to AES, similar in block size and key size options but using different mathematical methods than Rijdael. Both algorithms are harder to crack due to Moore's Law.

Key Topic!

Blowfish and Twofish

Blowfish and Twofish are two block cipher algorithms created by the renowned cryptographer Bruce Schneier. Blowfish operates on smaller data chunks (64 bits) and was designed for speed and efficiency. It was a popular choice in its time, but with computers becoming more powerful, its security has become less impressive.

Two fish is the successor to Blowfish, addressing the concern about data block size. It uses larger 128-bit blocks, offering stronger encryption. Both algorithms are based on a common structure called the Feistel network. While neither Blowfish nor Twofish have been proven to be broken, there are theoretical weaknesses. These weaknesses involve exploiting mathematical properties and how encryption keys are derived.

Summary of Symmetric Algorithms:

A few comparisons of the algorithms used thus far and their key sizes are shown in the Table below. The bit count of a cipher is its key size. Key strength or key length are other names for it.

Summary of Symmetric Algorithms

Algorithm Acronym	Full Name	Maximum/Typical Key Size
DES	Data Encryption Standard	56-bit
3DES	Triple DES	168-bit
RC4	Advanced Encryption Standard	256-bit
AES	Rivest Cipher version 4	128-bit typical
RC5	Rivest Cipher version 5	64-bit typical
RC6	Rivest Cipher version 6	256-bit typical
Twofish	Twofish	128-, 192-, 256-bit

RSA

RSA, a public key cryptography algorithm, is a secure protocol used in e-commerce and is slower than symmetric key algorithms. Unlike traditional methods where a single secret key is shared, RSA uses a unique key pair: a public key and a private key. The public key, as the name suggests, can be freely distributed. Anyone can use it to encrypt messages and send them to you. However, only your private key, which should be kept confidential, can decrypt these messages. This eliminates the need for pre-shared secret keys, a major challenge in traditional encryption. RSA goes beyond simple encryption. It allows you to digitally sign messages, acting like a virtual fingerprint that verifies your identity as the sender and ensures the message hasn't been altered. This feature is crucial for secure transactions and communication in e-commerce and protocols like TLS/SSL that secure websites.

The RSA algorithm uses integer factorization cryptography to multiply two prime numbers, derive a set of numbers, and create a private and public key pair. The private key decrypts encrypted data, allowing users to exchange information securely. For example, Alice sends Bob a message.

Bob can also prove his identity to Alice by, for instance, encrypting a digital certificate with his private key. Alice can decode it with his public key once she has it. The Table below provides an overview of these ideas.

Summary of RSA Public and Private Key Usage

Task	Which Person's Key to Use	What Kind of Key
Send an encrypted message	Receiver's	Public key
Decrypt an encrypted message	Receiver's	Private key
Send an encrypted signature	Sender's	Private key
Decrypt an encrypted signature	Sender's	Public key

RSA encryption is used in devices like SecurID USB dongles and hardware security modules (HSMs) for user authentication. It is vulnerable to man-in-the-middle attacks but can be defensible using digital certificates and PKI systems. Timing attacks can be defended using cryptographic blinding. To protect against attacks, a secure padding scheme is recommended, which includes random material and proof. Algorithm makers like RSA continuously release improved versions.

In 2000, RSA Security released the RSA algorithm to the public, free of licensing fees. They published PKCS standards to promote RSA techniques, including mathematical properties and HSM usage.

Diffi e-Hellman

The Diffi e-Hellman key exchange, invented in the 1970s, was the first practical method for establishing a shared secret key over unprotected communications channels.

Diffie-Hellman uses secure key exchange to establish a shared secret key for secret communications over a public network. Users Alice and Bob choose initial prime and base numbers, select secret integers, and compute equations to complete the shared secret. These static keys are discarded at the end of the session.

Diffi e-Hellman is a secure encryption algorithm used by the Transport Layer Security (TLS) protocol during encrypted web sessions. It works in ephemeral mode, generating keys for each key establishment process, achieving perfect forward secrecy (PFS). This ephemeral version, known as DHE or Ephemeral Diffi e-Hellman (EDH), requires more computational power. However, an elliptic curve alternative is available for a more efficient solution.

NOTE: Although the RSA method is significantly more widely utilized, the Diffi e-Hellman algorithm can also be employed within a public key infrastructure (PKI).

Elliptic Curve

Public-key cryptography offers another powerful tool besides RSA: Elliptic Curve Cryptography (ECC). Unlike RSA's reliance on complex number manipulation, ECC takes a unique approach based on elliptic curves – mathematical shapes resembling ellipses. This mathematical foundation allows ECC to achieve a significant advantage: smaller key sizes for the same level of security. Imagine carrying a tiny key ring instead of a bulky one – that's the benefit of ECC. Smaller keys translate to reduced processing power needed for encryption and decryption, making ECC more efficient for resource-constrained devices.

The world of ECC revolves around points on these elliptic curves. Specific curves and their properties are agreed upon by all parties involved. This paves the way for secure communication protocols. For instance, the well-known Diffie-Hellman key exchange algorithm can be adapted to work with elliptic curves in ECC (known as ECDH). This enables secure key establishment without the need for pre-shared secrets, a challenge in traditional methods.

ECC's efficiency and smaller key sizes have made it a popular choice for applications like smart cards, securing wireless connections, and communication channels like VoIP and IPSec. However, ECC is not without vulnerabilities. It's susceptible to specific attack methods and the theoretical threat posed by the future of quantum computing.

Quantum Cryptography

Quantum encryption, also known as quantum cryptography, is a reality that builds on quantum mechanics and quantum communications. It allows for the encoding of bits of information as quantum data, a method that would be impossible in classical digital encryption schemes. Currently, quantum cryptography is only a reality in the form of quantum key distribution (QKD), which uses a costly fiber channel to transmit quantum information. However, quantum encryption is not a perfect solution, as vulnerabilities are exposed over time.

More Encryption Types

The text highlights two additional encryption types: one-time pad and Pretty Good Privacy (PGP) application and encryption method, which are not included in other sections.

One-Time Pad

A one-time pad, also known as the Vernam cipher, is a stream cipher that encrypts plaintext with a secret random key, generating a random keystream. It uses the bitwise XOR operator to produce the ciphertext, making it difficult for attackers to guess the plaintext.

The pad encryption, a hand-calculated method, is known for its theoretically perfect security and is impossible to crack if used correctly. However, issues with this type of encryption have limited its widespread use, leading to the association of OTP with one-time passwords.

One-time pads, while having some drawbacks, are useful in secure environments for communication between users from different environments. They require perfect randomness, unlike computer-based random number generators, which are often pseudorandom number generators. The exchange of one-time pad data must match the message length, and proper disposal is difficult due to data remanence. They are also used in superencryption and quantum cryptography.

PGP

To try to make email conversations more secure, Pretty Good Privacy (PGP) is an encryption tool that is mostly used for signing, encrypting, and decrypting emails.

PGP, a hybrid cryptosystem, uses encryption specifications from the OpenPGP standard and offers a suite of tools for email, whole disk encryption, zip files, and instant messages. It uses a symmetric session key and asymmetric RSA for digital signatures and sending the session key, combining conventional systems and public key cryptography.

PGP uses 128-bit key sizes for data encryption, with newer versions allowing RSA or DSA key sizes from 512 to 2048 bits. The program uses hashing, data compression, symmetric key cryptography, and public key cryptography. Newer versions support OpenPGP and S/MIME, ensuring secure communications with various users.

PGP's security relies on the key size, with 2048-bit keys being safe against well-funded adversaries and supercomputers until 2020, while 1024-bit keys are strong enough for sensitive data environments, as they work with RSA.

Around the turn of the millennium, PGP creators and security-minded individuals developed OpenPGP, an open-source alternative to PGP. This standard allowed for easy integration with or replacement of PGP, leading to the development of GNU Privacy Guard (GPG), a combination of symmetric and public key encryption.

PGP, developed by Philip Zimmerman, is widely used globally for encrypting files before transit. Symantec owns the original PGP, which is available for free download, and various versions, including GNuPG, are available.

Pseudorandom Number Generators

Pseudorandom number generators (PRNGs) are used in cryptographic applications for unpredictable output. Developed within key generator programs, they use specific utilities like SHA2PRNG. To increase randomness, entropy is collected. PRNGs face threats from random number generator attacks, which can be prevented by randomness, AES, newer SHA versions, and system control.

AI and Genetic Algorithms

Artificial intelligence (AI) is getting smarter, and genetic algorithms are a fascinating way to boost its capabilities. Inspired by natural selection, these algorithms help identify individuals hidden within mountains of data.

Imagine searching for a specific person in a digital haystack. Genetic algorithms mimic evolution to achieve this. They start with a pool of data – text samples, code snippets, or financial records. Then, the algorithm selects potential matches and analyzes them based on specific criteria. In authorship identification, this might involve analyzing word choice, sentence structure, and vocabulary usage. Based on this analysis, the algorithm creates new "generations" of potential matches, combining features from promising candidates. This iterative process refines the search until individuals with a high likelihood of being the target are identified.

Stylometry, the analysis of writing styles, is a powerful tool within genetic algorithms. By analyzing factors like word choice, sentence structure, and word frequency, stylometry can act like a fingerprint, identifying anonymous works or narrowing down potential authors. This is just one example – genetic algorithms have a wider range of applications within AI. They can be used for statistical analysis, data mining to uncover hidden patterns, and even fraud detection by analyzing spending habits.

Hashing Basics

While encryption scrambles the message itself, there's another layer of security you can add: a hash. A hash is a summary of a message, often in numeric format, used in digital signatures, message authentication, and data protection. It is generated using a hash function to verify the message's integrity, typically after transit over a network. The function converts a variable-sized amount of data into a smaller block, producing a fixed-length hash value. The hash is created at the source and recalculated at the destination.

Hashs, also known as message digests, are condensed versions of a message, providing data integrity. They are one-way functions, easy to compute when generated but difficult to reverse. Hashs can be created without an algorithm, but typically require a cryptographic algorithm for reverse computation. They are similar to encryption methods that can be reversed.

Cryptographic Hash Functions

Cryptographic hash functions are like special mathematical black boxes. You feed them any piece of data, regardless of size, and they spit out a fixed-length string of characters, kind of like a unique fingerprint for that data. This fingerprint, called a hash value, is crucial for data security because it's nearly impossible to recreate the original data from just the hash, and very sensitive to any changes in the original data. This makes them ideal for verifying the integrity of information, like checking if a downloaded file is corrupt or ensuring a message hasn't been tampered with during transmission.

MD5

The Message-Digest algorithm 5 (MD5) is a widely used hashing algorithm designed by Ron Rivest, used to verify data integrity by checking the downloaded hash against the original hash. However, MD5 is susceptible to collisions, rainbow tables, and pre-image attacks, making it a potential threat. To protect against these attacks, use a stronger hashing function.

NOTE: A precomputed table called a rainbow table is used to decipher cryptographic hash functions. Passwords are most frequently cracked using rainbow tables.

SHA

One of the several hash functions created by the National Security Agency (NSA) of the United States and released by the NIST is the Secure Hash Algorithm (SHA).

The U.S. government uses SHA-1, a 160-bit hash, which has been deprecated in 2017. SHA-2 is more secure with 256-bit and 512-bit block sizes. Keccak, a 2012 SHA-3 winner, is not a replacement for SHA-2 as no compromise has been demonstrated.

Hashing algorithms must be collision-resistant to avoid the same output from two guessed inputs. Perfect hashing in cryptography is impossible due to unknowns like data and hash values. However, using a more powerful algorithm can increase collision resistance.

Government agencies began using SHA-2 in 2011 due to vulnerabilities in MD5 and SHA-1. Additional security involves comparing software keys to hardware keys for software activation.

RIPEMD and HMAC

RIPEMD, or RACE Integrity Primitives Evaluation Message Digest, is a 160-bit message digest algorithm used in cryptographic hashing, less commonly used than SHA and designed as an open-source hashing algorithm.

HMAC is a Hash-based Message Authentication Code (MAC) used to authenticate messages, ensuring integrity and authenticity by checking the cipher used and notifying the receiver of any modifications to the encrypted data.

HMAC calculates a MAC using cryptographic hash functions like MD5 or SHA-1, with corresponding MACs being HMAC-SHA1 or HMAC-SHA256, requiring careful selection of hash function type and version.

LANMAN, NTLM, and NTLMv2

Algorithms can also be used to hash passwords. While some password hashes are compromised because they have been broken, others are more secure than others. Beginning with the first hash, this section describes the Windows-based LANMAN, NTLM, and NTLMv2 hashes. Unless you are logging into a domain where Kerberos is the default authentication method, these three forms of authentication make an effort to safeguard your Windows PC login.

LANMAN

The initial hash used to store Windows passwords was called the LANMAN hash, sometimes referred to as the LAN Manager hash or just the LM hash. Although certain Windows versions support it in an effort to be backward compatible, it was a feature of Windows operating systems prior to Windows NT. The fact that the LM hash has multiple flaws and is readily cracked makes this backward compatibility potentially dangerous from a security standpoint. The LANMAN hash, based on the deprecated DES algorithm, has a maximum of 14 characters and can store only seven uppercase characters. This makes it vulnerable to brute-force attacks. It is recommended to disable the hash on operating systems running it by default and check for modifications.

The step-by-step instructions below demonstrate how to prevent Windows from storing LM hashes:

- **Step 1.** To see the Local Security Policy window, put secpol.msc at the Run prompt to open it.
- **Step 2.** Navigate to Local Policies > Security Options.
- **Step 3.** Double-click the Network Security: Do Not Store LAN Manager Hash Value on Next Password Change policy in the right pane.
- **Step 4.** As seen in Figure 14-3, select Enabled and then click OK. (Keep in mind that you are encouraging a negative in a circumstance like this.)

NOTE: It is necessary to use the Group Policy Editor rather than the Local Group Policy for Windows Server domain controllers. Normally, this would be completed at the default domain policy, but if needed, it may also be completed at the policy of a particular OU.

Modifying the Registry can disable the storage of LM hash passwords, especially for older Windows versions, as per Microsoft's "View Recommended Resources" document.

Microsoft recommends using a 15-character minimum password for storing LM hashes, as it cannot be cracked solely for user authentication. Disabling the LM hash policy is highly recommended due to potential policy limitations.

NTLM and NTLMv2

In the world of Windows logins, there's a history of authentication protocols, and NTLM (Network Trust Level Manager) is one of the key players. It all started with the LM hash, a method for storing passwords. However, the LM hash had a major limitation: it wasn't very secure and couldn't handle complex passwords.

This led to the development of the NTLM hash, introduced with Windows NT 3.1. It offered improvements like Unicode support for storing a wider range of characters in passwords and used the RC4 cipher for encryption. However, there was a catch – older systems still relied on the weaker LM hash for compatibility. To address these security concerns, Microsoft eventually disabled LM hashes by default in Windows Vista and Windows Server 2008.

Recognizing the limitations of NTLM, Microsoft introduced NTLMv2 with Windows NT 4.0 SP4. This was a significant upgrade, using a more robust 128-bit hashing algorithm that made it much harder for attackers to crack passwords. NTLMv2 ideally replaced Kerberos, another authentication system, as the default logon method in Microsoft domains. However, NTLMv2 still comes into play in situations where Kerberos is unavailable, users log in with local accounts, or the connecting device's operating system doesn't support Kerberos.

Hashing Attacks

Cracking hashes isn't always brute-force decryption by a supercomputer. While possible, attackers often take a more strategic approach. They might try creating collisions (two different messages with the same hash), exploit weaknesses during the hashing process (side-channel attacks), or leverage cracked hashes to gain higher access within a system (privilege escalation).

Let's talk about a few potential exploits for hashes.

Pass the Hash

An attacker can deceive a server or other system into believing they are authentic by obtaining the password hash of one or more user accounts and reapplying it. This technique is known as a pass-the-hash attack. The intention is for the attacker to obtain credentials belonging to a different user and perhaps elevate their privileges by breaking into the system, which is often a Windows server.

The attacker obtains hashes from a target system, collects user passwords using a hash dumping utility, and uses a "pass the hash" program to place the hashes within a server, like Windows Server's Local Security Authority Subsystem Service (LSASS). This attack can be carried out on individual client systems or Windows Servers, as they house numerous user account credentials.

To prevent such attacks, security measures like strong passwords, multi-factor authentication (beyond just passwords), and limiting user privileges (least privilege principle) are crucial. Additionally, restricting access to only trusted devices and implementing secure domain trusts can further tighten security.

Happy Birthday!

A birthday attack is a probability theory problem where two different messages with the same hash function are sent, causing a collision. The probability of this occurring with 366 people is 100%, based on a standard calendar year of 365 days.

The paradox of birthday attacks arises when fewer people are involved, with a 99% probability of a match and a 50% probability with only 23 people. To protect against birthday attacks, use strong collision resistance hashing functions and secure transmission mediums like SSH.

Additional Password Hashing Concepts

Quick Hits!

We've established that hashed passwords are the cornerstone of secure logins. But even with hashing, there are additional layers you can add for enhanced protection.

Attackers have various tools at their disposal to crack passwords. These include:

- **Guessing Common Passwords:** Hackers often try well-known passwords or dictionary words first.
- **Brute-Force Attacks:** They might systematically try every possible character combination, but this becomes time-consuming with longer passwords.
- **Rainbow Tables:** These are pre-computed databases that can help attackers quickly identify passwords from their hashes.
- Here's how to make password cracking much harder:

- **Strong Hashes:** Use robust hashing algorithms like bcrypt or scrypt. These take longer to compute, making brute-force attacks less feasible.
- **Key Stretching:** This technique transforms weak passwords into stronger, more secure ones, often increasing their size to make brute-forcing more difficult.
- **Salting:** Adding random data (salt) to the password before hashing makes it harder for attackers to pre-compute rainbow tables for common passwords. Even if they crack a hash, they would have to crack it again with the same salt for a specific user.

Always Remember!

There's a misconception that complex passwords are always better than longer ones. The truth depends on the attack method. Longer passwords, even simple ones, are generally more secure. However, strong policies should enforce both complexity and length. If enforcing long passwords isn't feasible, consider key stretching or salting as additional security measures.

- **Limit Login Attempts:** Restrict the number of consecutive password attempts to prevent automated attacks.
- **Multi-Factor Authentication (MFA):** Go beyond just passwords. Use OTPs (one-time passwords) like HOTP (HMAC-based OTP) or TOTP (time-based OTP) to add an extra layer of security.

CHAPTER 15:

ACRONYMS	*LIST OF ABBREVIATIONS*
PKI	*Public Key Infrastructure*
S/MIME	*Secure/Multipurpose Internet Mail Extensions*
SSL	*Secure Sockets Layer*
TLS	*Transport Layer Security*
PPTP	*Point-to-Point Tunneling Protocol*
PKI	*Public Key Infrastructure*
S/MIME	*Secure/Multipurpose Internet Mail Extensions*
L2TP	*Layer 2 Tunneling Protocol*
IPsec	*Internet Protocol Security*
CA	*Certificate Authority*
CSR	*Certificate Signing Request*
CRL	*Certificate Revocation List*
OCSP	*Online Certificate Status Protocol*
VPN	*Virtual Private Network*
SA	*Security Association*
AH	*Authentication Header*
ESP	*Encapsulating Security Payload*
HMAC	*Hash-based Message Authentication Code*
NGFW	*Next-Generation Firewall*
IDS/IPS	*Intrusion Detection System/Intrusion Prevention System*
BYOD	*Bring Your Own Device*
GNU	*GGNU's Not Unix*

Chapter 15

PKI and Encryption Protocols

The Security+ exam puts your cybersecurity knowledge to the test, and understanding encryption concepts is a critical area. This chapter dives deep into a powerful duo: Public Key Infrastructure (PKI) and Encryption Protocols. We'll explore how these technologies work together to secure connections, not just between users and servers, but across the entire digital landscape. From digital certificates verifying identities to robust encryption methods scrambling data, we'll explore the inner workings of PKI and its role in protocols like S/MIME, SSL/TLS, and SSH.

Quick Hits!

Public Key Infrastructure

A public key infrastructure (PKI) is a comprehensive system of hardware, software, policies, and people used to create, distribute, manage, store, and revoke digital certificates. It includes users, client computers, servers, services, and encryption. PKI is not the same as public key encryption, which is a method of achieving public key encryption. PKI creates asymmetric key pairs, a public key and a private key, which are kept secret and distributed as needed. Public key certificates, obtained by a web browser, are public and can be obtained by thousands of individuals, while private keys are not shared. PKIs bind public keys with user identities, often through certificates distributed by a certificate authority or a web of trust.

Certificates

Commonly Asked!

Certificates are digitally signed documents that bind a public key with a user's identity, often based on the X.509 standard, a common PKI standard developed by the ITU-T. These certificates can be used for connections to websites, email, and encryption. X.509 certificates include owner information and certificate authority information. They can be used internally or externally, but most people are familiar with SSL/TLS-based certificates used for secure HTTP connections. X.509 is the core of the PKIX working group, and more powerful hashing algorithms should be implemented.

SSL Certificate Types

Key Table!

SSL certificates play a crucial role in securing online communication. But with different types available, choosing the right one can be confusing. Companies classify certificates into domain, organizational, and extended validation certificates.

- **Domain Validation (DV):** This is the most basic and quickest to obtain. It verifies your control over a specific domain name but doesn't perform in-depth validation of your organization. Ideal for personal websites or blogs.
- **Organization Validation (OV):** Provides a step up in security by verifying your organization's existence and legitimacy. This involves a more thorough vetting process. Suitable for businesses wishing to establish trust with customers.

- **Extended Validation (EV):** Offers the highest level of validation, conducting the most rigorous checks on your organization's identity, legal status, and operational practices. EV certificates display a green address bar in some browsers, providing a strong visual cue of trustworthiness. Ideal for businesses handling sensitive data or financial transactions.

Single-Sided and Dual-Sided Certificates

Secure web sessions typically use single-sided certificates, where the server validates itself to users without requiring their identity validation. This is resource-intensive, especially for large servers with numerous connections. Dual-sided certificates, where both servers and users validate identities, are suitable for limited computer and session involvement.

Seen Mostly!

Certificate Chain of Trust

A certificate chain is a chain of trust where each component of a system is validated from the bottom up. The root certificate authority anchors this trust, followed by an end-entity certificate for a machine, handshakes with an intermediate certificate from a subordinate authority, and signs the intermediate certificate with its own private key. The root certificate authority employs code signing for integrity and authenticity. However, this can be defied, so security admins must be vigilant.

Note!

Certificate Formats

The ITU-T X.690 encoding formats include Basic Encoding Rules (BER), Canonical Encoding Rules (CER), and Distinguished Encoding Rules (DER). PEM, a Privacy-enhanced Electronic Mail format, uses base64-encoded ASCII files and can be identified by the.pem file extension. P12/PFX is a binary format based on PKCS#12 used to store server certificates, intermediate certificates, and private keys in one encryptable file..pfx, or Personal Information Exchange, is used by Microsoft for release signing. The.pfx file stores the certificate and its private and public keys..p7b format certificates can be used in IIS as the basis for S/MIME and single sign-on. Some extensions are also used for different types of data, such as private keys, and can be converted from one format to another using tools like OpenSSL.

Certificate Authorities

A certificate authority (CA) is a server that issues certificates to users, often used in Public Key Infrastructure (PKI) systems. CAs verify the identity of the recipient and negotiate the security of the connection between the user and the website. For a user to obtain a digital identity certificate from a CA, they must initiate a certificate signing request (CSR) and present two items of information: proof of the user's identity and a public key. The public key is then matched to the CA's private key, and if successful, the certificate is granted to the user.

To add security to the certificate validation process, SSL pinning or public key pinning can help detect and block many types of MIME attacks. It requires a secure and up-to-date web browser on each client. Recipients can use one or more certificates, with one-to-one mapping and many-to-one mapping being common. Registration authorities (RA) may be used to verify requests for certificates, but a CA is necessary if centrally storing certificates.

Certificate authorities are not just for the rich and famous; anyone can have a CA. Windows Server users can install their own CA using the Active Directory Certificate Services (AD CS) utility or randomly assign them.

However, CAs can be subverted through social engineering, and the certificate revocation list (CRL) is a list of no longer valid or revoked certificates. CRLs carry digital signatures to prevent DoS and spoofing attacks. An alternative to the CRL is the Online Certificate Status Protocol (OCSP), which contains less information and is less complex.

The Online Certificate Status Protocol (OCSP) is an alternative to the Certificate Revocation (CRL) in terms of security. It contains less information and is less complex, but does not require encryption. An alternative is OCSP stapling, which allows the presenter to bear the cost of providing OCSP responses. Key escrow is a secure copy of a user's private key in case of loss, preventing third parties from accessing encrypted communications. To avoid single points of failure, organizations can organize certificates in a hierarchical manner, with root CAs at the top and subordinate CAs below. An offline root CA can add a layer of security by not communicating with subordinate CAs or other computers. This approach requires secure policies, data loss prevention, and data loss prevention. It is important to be cautious during the planning stage of certificates to avoid exploits and vulnerabilities.

Web of Trust

A web of trust is a decentralized trust model that addresses public authentication issues in CA-based PKIs. It's peer-to-peer, with no root CA, and users can choose and share trusted certificates. This model can interoperate with standard CA architectures and allows users to self-publish their own public key information, increasing the likelihood of a certificate's legitimacy.

Security Protocols

Building on the trust established by PKI (Public Key Infrastructure), security protocols act as your digital bodyguards, encrypting communication and safeguarding data throughout your online interactions. Protocols like S/MIME and PGP scramble email content, while SSL/TLS secure connections for online transactions and logins. For remote access, SSH encrypts communication during login sessions, and PPTP/L2TP creates secure tunnels for virtual private networks. Understanding these protocols empowers you to navigate the digital world with confidence, knowing your data is protected by a layered security approach.

S/MIME

Secure/Multipurpose Internet Mail Extensions (S/MIME) is an IETF standard for cryptographic security in electronic messaging, including email. It uses a separate session key for each message and relies on PKI, X.509v3 certificates, and digital signatures for non-repudiation. S/MIME can be implemented in Outlook by obtaining a Digital ID and publishing it. However, it encrypts messages and potential malware, potentially compromising systems.

A must Know!

SSL/TLS

Secure Sockets Layer (SSL) and Transport Layer Security (TLS) are cryptographic protocols that provide secure Internet communications, including web browsing, instant messaging, email, and VoIP. They rely on a Public Key Infrastructure (PKI) for obtaining and validating certificates. TLS, a more secure solution, superseded SSLv3. The latest version of TLS is 1.2, with 1.3 as a working draft. SSL and TLS work similarly, requiring two types of keys: a public key and a session key. A recovery key is necessary if data is lost in an SSL/TLS session.

HTTPS, a combination of HTTP and SSL or TLS, is used in e-commerce and email protocols. It is important to note that SSL/TLS decryption and re-encryption can be a security risk and privacy issue, especially for BYOD users.

Another attack to watch for is the downgrade attack, which occurs when a protocol is downgraded from a high-quality mode to a low-quality mode or lower version. This is often implemented as part of an MITM attack. Preventive measures against downgrade attacks include using an IDS/IPS solution within the company network and using encrypted VPN tunnels for data sessions.

SSL can also be used by attackers, such as Trojans, who exploit the secure nature of SSL to exist undetected. Organizations can use next-generation firewalls (NGFWs) to filter out SSL-encrypted traffic in addition to regular firewalls used for unencrypted traffic.

SSH

Secure Shell (SSH) is a protocol that creates a secure channel between two computers or network devices, allowing remote control. Originally designed as a replacement for Telnet, it is widely used on Linux and Unix systems and Windows clients. SSH uses public key cryptography to authenticate remote computers and can be used for secure file transfers, copying, and tunneling. SSH 2.0 is incompatible with previous versions.

VPN Protocols for Secure and Private Connections

Virtual private networks (VPNs) are designed for secure, quick, and remote internet connections, utilizing PPTP and L2TP protocols with IPsec to ensure connection security.

PPTP

The Point-to-Point Tunneling Protocol (PPTP) is a protocol used in VPNs to encapsulate encrypted traffic. It is suitable for dial-up connections but not for VPNs without PPTP. PPTP can be used with CHAP-based authentication protocols, but MSCHAPv1 and MSCHAPv2 are vulnerable to dictionary attacks. EAP-TLS can dismiss these vulnerabilities, while L2TP with IPsec or tunneling protocols is recommended for environments with high session and data security.

L2TP

The Layer 2 Tunneling Protocol (L2TP) is a tunneling protocol used to connect VPNs, creating an unencrypted tunnel. It lacks confidentiality or encryption but is effective when paired with a security protocol like IPsec. L2TP also boasts improvements in later versions, with L2TPv3 offering enhanced encapsulation and security features. Setting up L2TP typically involves opening inbound port 1701 on servers and devices. Additionally, Windows Servers leverage PKI (Public Key Infrastructure) for added security within the L2TP framework.

IPsec

Internet Protocol Security (IPsec) is a protocol that authenticates and encrypts IP packets, securing communications between computers and devices.

It operates at the network layer of the OSI model and can negotiate cryptographic keys for mutual authentication. IPsec is made up of three other protocols: Security Association (SA), Authentication Header (AH), and Encapsulating Security Payload (ESP). It can be implemented in two ways: host-to-host transport mode, where the payload is encrypted but the header information is not, and tunnel mode, where the entire IP packet is encrypted and encapsulated. IPsec uses algorithms like SHA2 and possibly HMAC-SHA1 for integrity and authenticity, and can also use Triple DES and AES for confidentiality.

CHAPTER 16:

ACRONYMS	LIST OF ABBREVIATIONS
MTBF	*Mean Time Between Failures*
MTTF	*Mean Time To Failure*
MTTR	*Mean Time To Repair*
UPS	*Uninterruptible Power Supply*
MTBF	*Mean Time Between Failures*
NAS	*Network-Attached Storage*
SAN	*Storage Area Network*
RAID	*Redundant Array of Independent Disks*
SCSI	*Small Computer System Interface*
SATA	*Serial Advanced Technology Attachment*
BRI	*Basic Rate Interface*
PRI	*Primary Rate Interface*
DRP	*Disaster Recovery Plan*
FM-200	*DuPont FM-200*
RAID	*Redundant Array of Independent Disks*
CSU/DSU	*Channel Service Unit/Data Service Unit*
DNS	*Domain Name System*
FTP	*File Transfer Protocol*
PBX	*Private Branch Exchange*
SPS	*Standby Power Supply*
eSATA	*External Serial Advanced Technology Attachment*

Chapter 16

Network Redundancy for Maximum Uptime

In network design, ensuring consistent uptime is paramount. Downtime translates to lost productivity, frustrated users, and potential financial losses. Network redundancy serves as a critical strategy to combat these issues by introducing fault tolerance.

Key Topic!

Network redundancy involves incorporating multiple instances of critical network components and pathways. This can encompass redundant power supplies, network adapters, switches, and even internet connections. By employing these backups, the network can gracefully adapt in the event of a hardware failure. For instance, consider a network switch responsible for routing data packets between devices. Traditionally, a single switch failure could cripple network communication. However, with switch redundancy, a secondary switch seamlessly takes over, ensuring minimal disruption to data flow. Similarly, redundant power supplies and backup generators safeguard against power outages, preventing network-wide downtime. This multi-layered approach minimizes single points of failure and fosters a highly available network infrastructure.

Redundancy Planning

Note!

Let's face it, most networks could use a little more redundancy. You might be wondering why this keeps coming up. Here's the truth: I've seen countless network administrators overlook redundancy, leading to unnecessary downtime and headaches. Redundancy works – it's that simple!

This section dives into the world of redundancy planning, focusing on key areas: power, data, servers, internet connections (ISPs), and even physical locations (sites). By strategically planning for redundancy in these areas, you create a network that can weather most storms. Redundancy eliminates single points of failure, building a robust and resilient system.

Here's a breakdown of two common approaches to tackling single points of failure:

- **Redundancy:** This ensures uninterrupted system operation by having backups in place. Redundancy eliminates downtime but can be more expensive upfront. For example, you might have redundant power supplies or internet connections.
- **Spare Parts:** This method involves keeping spare parts on hand for critical components. While cost-effective, it requires taking servers or networks offline during repairs, leading to downtime. Spare parts are suitable for non-critical systems where downtime isn't a major concern.

Remember, the best solutions are often the simplest. Carefully evaluate your needs and resources to determine the ideal approach for your network. This is the situation (which we use for the remainder of this section on "Redundancy Planning"). You have the following powered equipment in your server room:

- Nine servers
- Two Microsoft domain controllers
- One DNS server

- Two file servers
- One database server
- Two web servers (which second as FTP servers)
- One mail server
- Five 48-port switches
- One master switch
- Three routers
- Two CSU/DSUs
- One PBX
- Two client workstations (for remote server access without having to work directly at the server);
- These are within the server room as well.

The server room already has redundancy, including two domain controllers, one with Active Directory, and two web servers for failover redundancy. The secondary system remains inactive until the primary fails, and two client workstations control the servers remotely. Redundancy is a detailed approach to preparing for potential system problems, ensuring the availability of servers and other equipment in a virtual server room. It involves envisioning various redundancy methods for each item.

Fail-open and fail-closed are two types of systems that can be used to ensure system continuity. Fail-open means that if a portion of a system fails, the rest of the system remains available, while fail-closed means that the entire system becomes inaccessible or shut down. The level of security required can lead to a mixture of fail-open and fail-closed systems. For example, a DNS server could fail-open to service the rest of the zones, while a database server might fail-close if a service or component fails. Redundancy and the level of security required also play a role in deciding which system to use.

Power is crucial for devices and computers, and it should be prioritized when planning for redundancy. Consider power implications like an engineer, enlisting engineering expertise or a third party for electrical requirements.

The server room is crucial for smart companies as it stores important data, settings, and apps. Power failure in this room could cause network downtime, resource loss, and damage to servers or devices.

When planning power redundancy, take a comprehensive approach, considering the needs of your entire network ecosystem. This goes beyond just the power demands of individual computers and servers. Factor in the additional power draw from networking equipment and Ethernet cabling. To create a layered defense against outages, consider redundant power supplies for critical systems. Additionally, deploy UPS devices to provide temporary backup power during short interruptions, allowing for graceful shutdowns. Don't forget the importance of your electrical infrastructure – ensure your circuits and electrical panel can handle the total power load. Maintaining clean, consistent power through voltage regulation is also essential. As a final layer of defense, consider backup generators to keep your network operational during extended power outages.

A solid preventative maintenance/security plan includes using the right power devices, which helps to safeguard a computer. You must defend yourself against many threats:

- **Surges:** An unanticipated rise in the voltage supplied is indicated as a surge in electrical power. This may be a gradual rise or a more significant rise referred to as a spike.
- **Spikes:** A spike is a brief voltage transient that may result from a tripped or short circuit.
- **Sags:** An unanticipated drop in the voltage supplied is known as a sag. Sags usually have a time limit and a voltage drop limit. But if the voltage drops much lower, a brownout can happen.
- **Brownouts:** A brownout is characterized by a voltage loss that usually results in dimming lights and computers turning out.
- **Blackouts:** A blackout is the complete absence of power for an extended length of time. A further issue related to blackouts is the potential increase that happens when power is restored. Lightning storms are the reason behind the spike in tech support calls that occur in the New York area every July! Systems are frequently damaged as a result of inadequate protection.
- **Failure of the power supply:** Power supplies and hard drives are similar in two aspects: They will, for one, fall short. It is a question of when rather than if. Two, when they start to fail, they may result in sporadic problems that are challenging to fix. Replace the power source if you think there may be a problem. Don't forget to use a backup power source.

Certain devices serve specific functions, while others offer multi-functional protection against various electrical problems. Now, let's discuss three of them: backup generators, uninterruptible power supplies, and redundant power supplies.

Redundant Power Supplies

A redundant power supply is an enclosure with two or more complete power supplies, ensuring that if one fails, the other takes over without computer failure. Common on servers, RAID boxes, and powerful workstations, it is essential to install redundant power supplies to all servers, switches, and routers, or consider scalable options.

Rearranged power supplies can be installed on servers to provide redundancy for devices. Manufacturers like HP, Thermaltake, and Enlight offer such systems, while Cisco offers AC power systems for networking devices. While this technology is useful in case of power supply failure, it doesn't protect against computer power disruptions.

Uninterruptible Power Supplies

Surge protectors are insufficient for protecting server room power issues. A UPS, which combines surge suppressor functionality with battery backup, protects servers from surges, spikes, sags, brownouts, and blackouts. UPS also acts as a line conditioner, cleaning up dirty power caused by noise, excessive devices, or power fluctuations. If a UPS doesn't solve the problem, contact your company's electrician for a solution.

Regularly test a separate line-conditioning device alongside a UPS, as line-conditioning devices supply power, while a UPS backup battery is only activated in case of power loss.

Emergency power is crucial for computer systems, ensuring they remain running for 5-30 minutes. UPS devices, with USB connections, communicate with the computer to shut down, suspend, or standby before the battery discharges completely. Most UPSs come with software for configuration.

The UPS battery's lifespan decreases with more devices connected, potentially leading to inconsistencies during power outages. Manufacturers limit battery backup-protected receptacles, and it's not recommended to connect a laser printer due to high current draw. The lead-acid battery requires 10-20 hours to recharge, and replacement requires a red light and a beeping sound. Overloading can also occur if power is no longer supplied.

UPS devices come in various levels, with standby UPS (SPS) being cheaper and having slight delays. Rack-mounted UPSs are full-blown, costing hundreds or thousands of dollars. Smaller devices, like SPSs, cost between $25 and $100. When planning for UPSs, consider the type and quality when data is crucial.

To guard against power outages, almost everything in the server room needs to be connected to a UPS (you'll probably need multiple). This covers everything in the server room, including workstations, security cameras, servers, monitors, switches, routers, CSUs and DSUs, PBX equipment, and more!

Backup Generators

If power fails in a building, a UPS can handle short outages, but not longer ones lasting hours or days. To maintain functioning, organizations need a backup generator to power lights, computers, phones, and security systems during short-term or longer outages.

A backup generator is a crucial component of an emergency power system, used to power a building during an outage, and should power the server room at least.

Backup generator fuel types include gasoline, diesel, natural gas, propane, and solar. Larger natural gas generators are commonly used for standby generators, which automatically start within seconds of a power outage. Companies should check company policy and municipal guidelines before planning and implementing a backup generator system.

Three categories exist for backup generators:

- **Portable gas-engine generator:** Gas-powered inverters are cost-effective, noisy, and require manual start, but are noisy, high-maintenance, and require extension cords. Gas-powered inverters are quieter but have higher prices.
- **Permanently installed generator:** Standby generators, typically natural gas or propane, are expensive, complex, and quieter, often connected directly to an organization's electrical panel, requiring minimal user interaction.
- **Battery-inverter generator:** Battery-inverter systems, based on lead-acid batteries, are quiet, user-friendly, and suitable for low-wattage environments or short power outages. They can be stored indoors but should be well-ventilated for fume release.

You should think about the following factors, among others while choosing a backup generator:

1. **Cost:** This will need to be budgeted for, just like any purchase made by an organization.

2. **How is the device started?** Does it turn on by itself? Most businesses demand this.

3. **Uptime:** How long will the generator run for before it needs to be filled with fuel? This is closely related to the bullet that follows.

4. **Power output:** What is the system's wattage? You should measure the whole maximum load your company might use at the main electrical panel by operating all computers, servers, lights, and other equipment concurrently before investing in a backup generator. As an alternative, you might calculate the sum on paper by combining the anticipated power needs of every gadget.

5. **Fuel source:** Does it run on gasoline, natural gas, or another fuel? The only fuel alternatives, if it's an autonomously starting system, will most likely be propane and natural gas.

Generac, Gillette, and Kohler are vendors offering backup generators. Companies often seek service contracts, which can be beneficial depending on the organization's size. Mission-critical devices like servers should constantly draw power from a line-conditioning device, with a UPS in case of power outage, and a backup generator if needed.

Redundant Data

Data failure can occur due to corruption, malicious intent, power failures, hard drive failures, and user error. To secure data, it should be redundant. Most servers should have redundant data sets, followed by databases and web servers. However, additional servers may benefit from failover systems instead of redundant data arrays. RAID is not necessary for most client computers' data.

RAID offers two key benefits: enhanced data protection and potentially faster data access speeds (depending on the RAID level chosen). RAID arrays can be internal or external, and can be SCSI, SATA, eSATA, or Fibre Channel. The level of RAID depends on the number of disks and their working together. For the exam, understanding RAID levels, including RAID 0, RAID 1, RAID 5, RAID 6, and RAID 10 (RAID 1+0), is essential.

RAID Descriptions

RAID Level	Description	Fault Tolerant ?	Minimum Number of Disks
RAID 0	Striping To improve performance, data is striped over several disks.	No	Two
RAID 1	Mirroring Two identical disks are used to copy the data. The other disk keeps working even if one fails. Figure 16-1 provides an example of this. Because there is a full copy of the data available at all times, this RAID variant permits the least amount of downtime. Disk duplexing is the process of connecting every disk to a different controller.	Yes	Two (and two only)
RAID 5	Striping with Parity Multiple disks are used to strip data, and each disk is also written with fault-tolerant parity data. The array can use the parity information to reassemble the data in the event that one disk fails. Figure 16-2 provides an example of this.	Yes	Three

RAID 6	Striping with Double Parity Similar to RAID 5, data is striped over several disks, but parity information is striped twice. Usually, this calls for an additional disk in the array. This solution is more suitable for time-sensitive applications and can continue to function even in the event of two failed drives.	Yes	Four
RAID 0+1	Brings together the benefits of RAID 0 and RAID 1. requires four disks at the very least. There are two RAID 0 striped sets in this system. The two sets are the same.	Yes	Four
RAID 10 (also known as 1+0)	Combines the benefits of both RAID 0 and 1. usually needs four disks at the very least. At least two RAID 1 mirrors that are striped are present in this system.	Yes	Four

RAID 1 and RAID 5 are two types of arrays that distribute data blocks and parity information. RAID 1 distributes data to both disks, known as the M: drive or M: volume. In RAID 5, blocks of data are distributed and parity information is written to each disk. If one disk fails, parity information from other disks reconstructs the data. Fault tolerance allows hard drives to function without downtime, while backup involves copying data for archival.

Windows servers support RAID 0, 1, and 5 within the operating system, but most client operating systems cannot. Hardware controllers and some motherboards have built-in RAID functionality. Hardware is better for RAID control than software within an operating system. Network-attached storage (NAS) points can be combined to form a storage area network (SAN) but costs more for organizations.

RAID can be categorized in three different ways. Knowing these classifications will help you decide what kind of RAID system to put in place.

1. **Failure-Resistant Disk Systems:**

This category prioritizes protecting your data against **drive failure**. RAID 1 (mirroring) is the champion here. It creates an exact copy of your data on another drive. So, if one drive fails, your data remains safe and accessible on the mirrored drive. This offers the highest level of data security within a single storage system, but it also comes at the cost of using double the storage space.

2. **Failure-Tolerant Disk Systems:**

This category focuses on ensuring your system can withstand the failure of **a single component**, not just a drive. RAID 1 mirroring (including duplexing) falls under this category as well. However, other RAID levels like RAID 5 and 6 also play a role here. These RAID types distribute data and parity information across multiple drives, allowing the system to tolerate a single drive failure without data loss.

3. **Disaster-Tolerant Disk Systems:**

This category takes things a step further by aiming to protect your data against even larger disasters. Here, the focus is on physical separation. RAID 0+1 (a combination of striping and mirroring) is an example. It creates redundancy not just within a single storage system, but across physically separate locations. This way, even if a fire or flood destroys one location, your data remains safe in the other.

The data must be accessible and highly available, with a well-planned RAI D system having high availability and scalability. Some systems have better elasticity, allowing for dynamic hot-swappable disks and space addition. Some data may be moved to the cloud for additional elasticity.

Redundant Networking

Quick Hits!

Network connections can also break. As everyone knows, users must have the network operational at all times, or else there will be severe consequences. Networking connection failures might jeopardize an organization's security. The following are a few connection kinds to think about:

- Server network adapter connections
- Main connections to switches and routers
- The Internet connection

Basically, any highly significant network connection that has the potential to fail is what I mean when I talk about redundant networking. These connections are usually found in the server room.

Redundant network adapters are used to reduce server downtime when one fails. They can be set up using applications or switches that control data direction. The choice depends on company policy, budgeting, and equipment. When installing multiple adapters, the computer becomes a multihomed machine, and it's important to consider their normal behavior during failure. Some options include multiple physical adapters or a single card with multiple ports, such as a multi-Ethernet port Intel network adapter. This provides a single point of failure.

Companies should have at least one backup switch to avoid single points of failure. Star-bus configurations can be single points unless special backup ports are used. Hierarchical stars offer more security due to redundant topology, allowing certain areas to function even if one switch fails, preventing network collapse.

Organizations often rely on one Internet connection for their entire network, which can lead to a single point of failure. To prevent this, consider secondary connections like redundant fiber optics, BRI, PRI, or dial-up connections. Some companies also install fault-tolerant, dual Internet connections. If using a web host, consider having multiple mirror sites or more. If an organization is without internet for more than a day, network and security admins will be the first to be affected.

NOTE: Devices on networks will malfunction. All that needs to happen is time. The idea of mean time between failures (MTBF), a reliability phrase that provides an average number of device failures per million hours of usage, was discussed in Chapter 12, "Vulnerability and Risk Assessment." When thinking about redundant networking, you should also take mean time to failure (MTTF), mean time to repair (MTTR) and mean time to failure (MTBF) into consideration.

Redundant Servers

Clustering multiple servers together, even if redundant network adapters and disks are insufficient, can provide low downtime and a secure feeling, despite being more costly and requiring more administration.

Quick Hits!

There are two different categories of server clustering:

➔ **Failover Clusters:** High-availability clusters are designed to allow a secondary server to take over if the primary server fails, reducing the risk of a single point of failure. Examples include using two Microsoft domain controllers, tertiary and quaternary servers, and a secondary DNS server for fail-closed systems. These clusters can be used to reduce the number of servers that might fail concurrently.

➔ **Load-balancing Clusters:** Load-balancing clusters are connected computers that share resources like CPU, RAM, and hard disks, enabling parallel, high-performance computing. Microsoft's Cluster Server is an example, and third-party vendors offer clustering software for operating systems and virtual operating systems.

Data replication between servers, often with database and web servers, is a combination of redundant data and server clustering. However, local servers may be affected by attacks or disasters.

Always Remember!

Redundant Sites

Redundant sites, a mirror of the entire network, are essential for disaster recovery. They act as a safe haven for data and users, balancing redundancy and disaster recovery methods. The better the redundant site, the less time lost and the less perceived disaster it is, depending on the organization's chosen type.

A must know!

Organizations typically choose warm redundant sites, as compared to the three bears' bowls of porridge, as they are more suitable for their needs.

- **Hot Site:** Hot sites are a quick-replicated version of an organization's original site, enabling full recovery in case of a disaster. They include computer and phone installations, a simulated server room, and regular data replication. These sites are crucial for companies facing financial ruin.
- **Warm Site:** Warm sites have computers, phones, and servers, but require security before users can work. They have backups and are preferred by organizations for cost-effectiveness.
- **Cold Site:** A non-productive site requires tables, chairs, bathrooms, and technical setup, but requires computer and data restoration for proper use, suitable for companies able to handle week-long non-productive periods.

Backup sites, redundant but likely to be required in disaster situations, are crucial for network security administrators to plan for redundancy and fault tolerance before resorting to disaster recovery methods.

Redundant People

A must Know!

The redundancy of a person's role in an organization is a significant issue, with IT management personnel typically having an average employment length of less than five years. This attrition is largely due to individuals moving to other departments, leaving, or retiring. To address this, succession planning is crucial, identifying internal IT infrastructure experts who can take over if a key decision-maker departs.

A must Know!

Disaster Recovery Planning and Procedures

Disasters come in all shapes and sizes, from natural disasters like floods and fires to cyberattacks and unexpected outages. Network security administrators need a proactive approach to ensure data security, server uptime, and business continuity in the face of such threats. A robust disaster recovery plan (DRP) is your network's shield, focusing on three pillars: data protection with regular backups to secure off-site locations, server room security with physical safeguards, environmental controls, and redundant power, and finally, site redundancy through geographically dispersed backup sites or cloud solutions to ensure operations continue even if your primary location is unavailable.

A must Know!

Data Backup

Disaster recovery (DR) is crucial for data backups, tailored to an organization's needs and budget. Data can be backed up to various media, computers, SANs, NAS devices, or the cloud, with tape backup being the best local medium.

Understanding tape backup types is crucial for exams, as most operating systems and third-party utilities support these types. Selecting the backup type, whether it's a folder or an entire volume, is essential for simplicity.

- **Full Backup:** Full backups store all folder contents on one or more tapes, but can be expensive and time-consuming due to space and data usage.

Incremental and differential backups are often used alongside full backups in a backup plan.

- **Incremental Backup:** Incremental backups back up only the contents of a folder that has changed since the last full or incremental backup, and must be preceded by a full backup. Restoring a folder or volume requires starting with the full backup tape and moving to incremental backups chronologically.
- **Differential Backup:** Differential backups only back up folder contents that have changed since the last full backup, and must be preceded by a full backup. Incremental backups don't recognize differential backups.

An example of a basic one-week backup plan utilizing the full and incremental backup types is provided in the Table below. On Mondays, a complete backup is completed, and incremental backups are completed Tuesday through Friday.

A must Know!

Example Incremental Backup Schedule

Day	**Backup Type**	**Time**
Monday	Full backup	6 p.m.
Tuesday	Incremental backup	6 p.m.
Wednesday	Incremental backup	6 p.m.
Thursday	Incremental backup	6 p.m.
Friday	Incremental backup	6 p.m.

The schedule requires five daily backup tapes, with a dummy tape replaced every morning to prevent data theft. Real tapes are locked up until needed. Organizations can archive specific tapes weekly, use new ones every Monday, or run a complete full backup over the weekend and archive it every Monday, as long as no data loss is reported.

A server was backed up using a backup procedure. If the server crashed, a backup operator would review logs to determine the crash date. After an admin fixates the server, the operator would restore the data using three tapes, starting with Monday's full backup tape.

Another potential backup schedule is displayed in the Table below, with a complete backup performed on Monday and differential backups performed on Wednesday and Friday.

Example Differential Backup Schedule

Day	**Backup Type**	**Time**
Monday	Full backup	6 p.m.
Tuesday	None	
Wednesday	Differential backup	6 p.m.
Thursday	None	
Friday	Differential backup	6 p.m.

The backup operator needs two tapes for data restoration due to a weekend failure. The differential backup backed up everything since the last full backup, while the incremental backup schedule requires each tape for restoration. Differential backups may back up items from previous backups, while incremental backups use the clear archive bit.

Windows Server manages full and incremental backups for administrators, with 2008 R2 and higher creating full backups that can be recovered, occupying only necessary space.

The schedules that we just displayed in Tables 16-2 and 16-3 are fundamental backup techniques, which are also referred to as backup rotation systems. Similar actions could likewise be taken by organizations during a two-week period. You should be aware of a few additional backup plans, though, as they are also utilized in the field. These could make use of one or more of the previously listed backup formats.

- **10 Tape Rotation:** This method allows easy access to backed-up data during a two-week backup period, using each tape daily for two weeks. The entire set is recycled, with differential backups on Mondays and full backups on Fridays. Choose the best option based on the required tapes and time.
- **Grandfather-Father-Son:** The backup rotation scheme is a common method involving three sets of tapes, corresponding to son, father, and grandfather. These tapes are rotated daily, weekly, and monthly, with the last week's backup being transferred to father status.
- **Towers of Hanoi:** The backup rotation scheme, based on the Towers of Hanoi puzzle, uses three different backup sets. The first tape is used every second day, the second every fourth day, and the third every eighth day. This scheme can be extended to include additional tapes, but it becomes more complex to remember the order.

Example of Towers of Hanoi Three-Tape Schedule

Day of the cycle								
	1	2	3	4	5	6	7	8
Tape	A		A		A		A	
		B				B		
				C				C

To prevent data rewriting, start on the fourth day of the cycle with tape C and plan during the planning stage. Test a restore sequence and store tapes in a cool, dry area away from sunlight and power sources. Test tape backup methods and integrity by restoring parts. Archive data to a third party for backup or full replication.

Backing up groups of files can also involve a snapshot backup, which backs up an entire system or architectural instance. This is typically done when a new system is installed or major changes occur. Some organizations even backup system images monthly or weekly, requiring adequate IT budget planning.

Organizations often back up to tape, but in "big data" scenarios, data might be stored on the cloud or archived with a third party. Ensure archival offsite, geographically distant from the main site, as part of disaster recovery planning.

DR Planning

Before we can make plans for catastrophes, we must define exactly what types of disasters are possible and rank them from most likely to least likely. It may sound a little gloomy, but doing this is essential to guaranteeing the long-term viability of your company.

How could something go wrong? As our scenario, let's concentrate on the server room at the start of the chapter. We have nine servers, networking gear, a PBX, and a few workstations, as you recall—a fairly standard server room for a midsized business. Remember that when it comes to disaster recovery planning, larger firms will have larger server rooms, more equipment, and other factors to take into account.

There are two types of disasters: man-made and natural. The following are a few of the calamities that could render your server room unusable:

- **Fire:** Fire is the most common planned disaster due to the need for fire suppression systems in municipalities and organizations. The three main types of fire extinguishers are A, B, and C. However, these systems are not adequate for server rooms, as they can damage equipment. A server room should have its own fire suppression system, such as DuPont FM-200, which uses a large tank and nozzles to quickly extinguish fires. Regular fire suppression alarm tests and drills are essential to ensure personnel know how to respond to emergencies.
- **Flood:** To prevent flood damage to server rooms, it's crucial to place them on the first floor or higher, away from boiler rooms and other water sources. The room should be three-dimensionally analyzed and protected, including floors, walls, and ceilings. Some server rooms may be designed to be a room within a room with drainage installed.
- **Long-term power loss:** A backup generator and possibly a redundant site are necessary for long-term power outages, although the UPS should be able to handle short-term power outages.
- **Theft and malicious attack:** Physical security measures like door locks and video cameras are crucial to prevent data theft and malicious attacks. Servers should be cable-locked and removable hard drives have key access. A security administrator is responsible for writing policies and enforcing them, including hiring security guards and terminating employees. Chapter 7 covers malicious network attacks and their prevention.
- **Loss of building:** Temporary building loss due to a gas leak, attack, crime scene investigation, or natural event necessitates personnel accessing a

redundant site. Server room data archived, warm site maintained, and plan for data restoration and network functionalization.

Disaster recovery plans (DRPs) should not include information about the day-to-day operations of a company, such as computer updates, patch management, monitoring and audits, and so forth. Instead, they should contain information about redundancy, such as sites and backup. A disaster recovery plan should only contain the information that is absolutely required. When a calamity does occur, having too much information can make it difficult to utilize.

Your disaster recovery plan should include the following documented policies, processes, and information, however, they are not all-inclusive:

- **Contact Details:** How staff members will get in touch with the organization and who you should call in the event of a tragedy.
- **Impact Determination:** A process to ascertain the entire extent of an organization's impact following a tragedy.

This involves calculating the cost of replacing lost assets and evaluating them.

- **Recovery Plan:** This will depend on how the impact of the tragedy is determined. Several variations exist based on the kind of calamity. While it is not feasible to anticipate every scenario that can arise, the aforementioned inventory provides a useful foundation. The recovery plan outlines a series of phases that specify what will be recovered when and in what order, along with an estimated time frame for recovery. An after-action report (AAR), a formal document created to assess a recovery plan's efficacy in the event that it is put into practice, may also be included.

- **Business Continuity Plan:** A Business Continuity Plan (BCP) is a crucial plan for businesses to continue operations in case of a disaster. It includes business impact analysis recovery time objective (RTO) and recovery point objective (RPO). By using proper archival, hot/warm/cold sites, and redundant systems, organizations can determine a general timeframe for data recovery, forming part of IT contingency planning (ITCP). It is often carried out by a team.

Organizations have a continuity of operation planning or crisis management group that conducts tabletop exercises to discuss the Business Critical Control Point (BCP). This method saves time and reduces disruption, helps identify critical systems, mission-essential functions, failover functionality, and alternate processing sites, and assesses potential disaster impacts.

- **Agreement Copies:** The DR plan should be kept in the same file as any agreements you may have with building management, ISPs, redundant site vendors, and so forth. In "Policies and Procedures," Chapter 18, we talk about agreements.
- **Disaster Recovery Exercises and Drills:** Workers should receive instruction on what to do in the event of a disaster. These activities must follow safety regulations and be meticulously documented.
- **Hierarchical List of Critical Systems and Critical Data:** This is a list of mission-essential data and systems for business operations, prioritized by priority. It should include geographic considerations like offsite backups, physical distance, and legal implications like data sovereignty. Client computers, test computers, and training systems should be last or not listed.

The CSO-(Chief security officer) or high-level executive typically handles disaster recovery planning, often with ISSO assistance. The size and management type of the organization determines the benefits. Proper DR planning benefits any size organization, and compliance rules should be consulted. Other organizational factors may also be considered.

CHAPTER 17:

ACRONYMS	LIST OF ABBREVIATIONS
OSINT	*Open Source Intelligence*
PII	*Personally Identifiable Information*
IT	*Information Technology*
SCADA	*Supervisory Control and Data Acquisition*
RFI	*Radio Frequency Interference*
EMI	*Electromagnetic Interference*
TEMPEST	*Transient Electromagnetic Pulse Emanations Standard*
CAN	*Controller Area Network*
UAV	*Unmanned Aerial Vehicle*
GPS	*Global Positioning System*
CA	*Certificate Authority*
Wi-Fi	*Wireless Fidelity*

CHAPTER 17

Examining people and their conduct as well as the resources available to your organization is the premise underlying this chapter.

This chapter discusses protecting employees' privacy while protecting organizational infrastructure from social engineers and malicious individuals. Policies and procedures can safeguard legitimate individuals and infrastructure.

An organization's facilities, including buildings, environmental controls, vehicles, and other assets, must be protected to ensure security. Fire suppression systems, shielding, and electrical systems should be implemented consistently.

This chapter covers technology security concepts, starting with social engineering, user education, and the organization's building and facilities. It emphasizes the importance of preventing social engineering attacks, increasing user knowledge, and addressing the connection between technology implementation, user interaction, and the potential for con artists.

Always Remember!

Social Engineering

Information security-based social engineering is a low-level yet effective form of behavior, affecting 1 out of 10 people annually, with half unaware of its existence. This can have devastating consequences for organizations and innocent individuals.

Chapter 1 of "Introduction to Security" explains that social engineering involves manipulating users to reveal confidential information or perform harmful actions. Examples include phone calls for usernames and passwords, and malicious individuals use them to steal valuable assets.

Experts in social engineering employ strategies and ideas like the following:

- Authority
- Intimidation
- Bold impersonation
- Urgency, scarcity, and even emergency
- The grooming of trust/familiarity/liking
- Persistence and patience
- Relating to the user: using company jargon, consensus/popular decision, and social facts and
- proof
- Embedding of questions within conversations

Social engineers exploit information sources like open source intelligence (OSINT) and social networking sites to gain knowledge about targets. They use tools like P2P software to obtain information disclosure. Social engineering success is due to a lack of user awareness and can be effective in environments with little training or public areas. Common types include sharing office space.

Pretexting

Pretexting involves creating a scenario to trick a victim into sharing information. It often involves impersonation to obtain records about an organization. IT employees should be vigilant for impersonators and seek identification. If doubts arise, escalate to supervisors or authorities.

Malicious Insider

Malicious insider threats are insidious attacks where individuals impersonate personnel to gain access to secure data, PII, financials, and engineering plans. These attacks are often part of corporate espionage plans and can be initiated by organizations from other countries. Companies often conduct thorough background checks, credit checks, and psychological assessments before hiring individuals, with a trial period allowing limited access to secure data and environments.

Malicious insiders can be devices or bugs inserted into organizations using social engineering skills, requiring physical access to the building, making identification and authentication paramount.

This book highlights the underappreciation of malicious insider threats by organizations. They have a better chance of obtaining desired results than outsiders. For example, stealing 100,000 credit card numbers could be done through MITM attacks, hacking through bank firewalls, or getting inside. The potential for success outweighs the risk, with numerous compromises to banks and stores.

Diversion Theft

Diversion theft involves a thief stealing millions of dollars-worth of IT equipment daily by diverting delivery to a nearby location. Couriers and shippers must know their destination and the organization's contact information.

Phishing

Phishing is the fraudulent act of obtaining private information through electronic communication, often disguised as someone else. It involves targeting thousands of individuals without much background information. An example is an email requesting information verification, leading to a malicious website that appears legitimate. Phishing techniques often involve posing as a vendor.

A fake order can lead to credit card fraud, ID theft, and even Trojans on the bogus web page. The person may click on the link to track the order, leading to the fraudulent web page. This cyber-bullying can be a significant issue, highlighting the need for careful handling of such orders.

Phishing attacks typically require no target information, but some attackers target specific groups or individuals, known as spear phishing, or senior executives through whaling attacks, which require extensive information about the target, often available online.

Phone phishing, also known as vishing, is a scam that involves a prerecorded phone call from a legitimate institution, asking for confidential information under the guise of verifying it for protection. Vishing can be easily perpetuated on large groups of people using automated systems, such as telemarketers.

***NOTE*:** War-dialing is a comparable automated system technique. To find computers and fax machines, a device (modem or other system) is used to scan a list of phone numbers and dial them. The method separates the phone numbers linked to computers from the ones linked to voice lines. It produces a list that other attackers can utilize later on for other objectives.

Phishing, a type of social engineering, typically involves email and websites for private information. To protect against phishing, install a phishing filter or add-on on your web browser, and be aware that institutions don't request private information. If unsure, hang up or delete the email. Hover over a link to identify a different URL. Many phishers use spyware applications, enabling additional phishing attempts beyond the initial website.

Hoaxes

A hoax is a deceptive attempt to deceive people into believing false information, often carried out in person or through communication methods, while phishing is primarily confined to e-communication and phones. Hoaxes can consume valuable organizational resources, such as email replies and internet bandwidth. Examples include Google's name change and Bill Gates' assassination. Companies may set time limits for hoaxes to indicate non-productive behavior.

Con artists use deceptive tactics like pretexting, phishing, hoaxes, malicious insider attempts, and diversion stealing, which are together referred to as confidence tricks. Even less secure approaches exist, though, and they are frequently combined with the earlier techniques to obtain people's personal information. These comprise piggybacking, dumpster diving, eavesdropping, shoulder surfing, and baiting.

A must Know!

Shoulder Surfing

Shoulder surfing is a form of no-tech hacking where a person uses direct observation to steal a target's password or authentication information. To prevent this, users should shield devices, lock computers when not in the immediate area, and implement password masking. However, lesser devices may not implement masking by default, potentially violating company policies.

A must Know!

Eavesdropping

Eavesdropping is the act of directly observing someone to "listen" in on a discussion. A person could be listening in on a phone conversation or lurking around a corner. To prevent eavesdropping, soundproof rooms, and encrypted phone conversations are common solutions.

Dumpster Diving

Dumpster diving is the act of physically searching through trash and recycling bins for personal information. For as long as feasible, any critical documents should be kept in a secure location.

They ought to be destroyed after they are no longer needed. (A few businesses burn their paperwork.) Not only can information be discovered on paper, but it can also be found on removable media or hard drives. Hard drive disposal and/or recycling are addressed later in this chapter.

Baiting

Baiting is the act of a malevolent person leaving removable media, like an optical disc or USB drive, lying around in the open with malware on it. It may have a unique appearance or an intriguing logo. The malware infects the computer when someone connects it to his computer, attempting to take over both the computer and the network it is a part of.

A must Know!

Piggybacking/Tailgating

Piggybacking and tailgating are security threats where unauthorized individuals tag along with authorized individuals to gain access to restricted areas. Mantraps, small spaces with interlocking doors, can be used to defeat these attacks. Multifactor authentication is often used with mantraps, while turnstiles, double entry doors, and security guards are less expensive and less effective solutions.

Watering Hole Attack

The watering hole attack targets users by loading malware on common websites, hoping they will access them and activate the malware, potentially infecting their system and spreading through the network, potentially building upon other social engineering methods.

Malware attacks target smaller websites, such as those used by companies manufacturing widgets. These websites are often known for their lack of security and are easily infected by attackers. If many users visit these websites, malware can be installed and spread throughout the company. Additionally, attackers may redirect users to other sites with more hardcore malware, such as ransomware.

As a security administrator, you cannot force companies to implement malware prevention methods like software patches and secure coding. Instead, focus on localized prevention methods like user training, browser reduction, blacklisting, and monitoring.

Summary of Social Engineering Types:

A must know!

Type	Description
Pretexting	When someone fabricates a situation, or pretext, to get a victim to provide information.
Malicious insider threat	When an employee works for a company with the intent to gather PII, financial information, design work, and secret information.
Diversion theft	When a thief attempts to take responsibility for a shipment by diverting the delivery to a nearby location.
Phishing	The endeavor, usually carried out electronically, is to collect personal information falsely. The phone is used for vishing. Spear phishing focuses on particular people. Whaling goes after high-ranking officials.
Hoax	The endeavor to mislead others into accepting a false belief.
Shoulder surfing	When someone deciphers a target's password, PIN, or other authentication information by direct observation.
Eavesdropping	When someone "listens" in on a conversation through direct observation. This might be someone listening in on a phone conversation or someone lurking around the corner.
Dumpster diving	When someone raids recycling and trash bins looking for personal information.
Baiting	When a malicious person purposefully leaves removable media that is infected with malware, like an optical disc or USB drive, out in the open, hoping that someone will bring it back to their computer and use it.
Piggybacking/tailgating	When someone who is not authorized enters a restricted place by tagging along with someone who is.
Watering hole attack	When a hacker targets users' unique browsing preferences in the hopes that they would visit specific websites and activate malware that is concealed there.

Social engineering can sometimes be a simpler method of information theft than other, more sophisticated techniques.

For instance, it might be much simpler to fool someone into disclosing their password than it would be to try to crack it if someone with evil intent wanted to know it.

User Education

User resources are highly exploitable for companies, as they are not like computers or robots. Human error remains a potential threat. Education is the best method to prevent social engineering attacks, malware infection, and user error.

As an IT manager, I prioritize technology and security training for employees, using various platforms for accessibility and entertainment, including classrooms, digital courses, and fun posters in the cafeteria.

User training faces several obstacles, including organizational acceptance, employee involvement, and creating a reward system. Executives are increasingly adopting user education as a standard practice. To counter against resistance, counter with news articles and case studies. Employees can also be trained themselves, using incentives or penalties. Some organizations hire IT trainers or offer incentives for attendance, but incentives often outperform penalties.

Role-playing can be a fun and effective way to teach employees about various organizational roles, such as system administrators, privileged users, and data owners. By creating scenarios and quizzing employees, they can expand their knowledge and develop solutions, ultimately enhancing the organization's security and strength. This approach can be akin to a table-top job rotation.

Time constraints often prevent individuals from dedicating sufficient time for training. To save time and enhance knowledge, a company-focused mindset and human resources can be fostered, focusing on efficiency and benefits for both the company and the individual.

That being said, the content of the training is more crucial for the Security+ certification than the method of completion. When educating staff members, you can cover the following fundamental rules:

- Never, ever divulge any authentication information, including passwords, PINs,smart cards, tokens, company IDs, and so forth.
- Protect keypads and screens at all times when entering login data.
- Comply with the company's clean desk policy, which stipulates that any paperwork, devices, and personallyowned gadgets, along with additional stuff, be stored (or secured) when the user isn't at his or another workspace, like her desk.
- Consistently monitor your phone calls and emails, and record everything that happens. Another name for this is communications screening.
- When it's feasible, encrypt data, phone conversations, and emails.
- Record the circumstances and report it to your supervisor, security, or the appropriate authorities if you have any reason to dispute the legitimacy of a person, email, or phone conversation.
- Never handle or utilize any unfamiliar removable media.
- Any sensitive material that is going to be thrown out or recycled should always be shredded.

- When it comes to the management and disposal of data, always abide by business policy. For instance, make careful to dispose of a hard drive, USB flash drive, memory stick, or optical disc correctly if it is no longer in use. If the user is unsure, they should get in touch with the organization's facilities or IT departments to determine if the item should be disposed of or recycled.
- Consistently monitor and expedite shipments.
- Use caution at all times when using a web browser. Before clicking Go or Enter, review everything you typed twice. Nothing should be clicked until you are certain of what it is. I gave this one three stars since a significant portion of organizational harm is caused by web-based attacks.

When training employees, keep them interested and incorporate fun and humor. Avoid giving out authentication details if it's okay for them. Provide real examples of social engineering to connect them to their defenses. Social engineers exploit weaknesses in human nature, regardless of an organization's firewall or armed guards, ensuring the health of the organization.

The list of social engineering methods and defenses is extensive, but policies and procedures, along with user education and awareness, are effective tools against such threats.

Facilities Security

IT directors and building management are responsible for technology facility installation, maintenance, and repair. Basic knowledge of environmental controls like fire suppression, HVAC, equipment shielding, and company vehicles is crucial, with fire suppression being the most important concept.

Fire Suppression

Fire suppression is the control and/or extinguishing of fires to protect an organization's employees, data, and equipment. There are three types: handheld fire extinguisher solutions, sprinkler systems, and special hazard protection systems used in server rooms.

Fire Extinguishers

Select a handheld fire extinguisher carefully, as they can cause damage to electronics and should not be used in critical areas like server rooms or server rooms.

Some classifications of fires and their related fire extinguisher signs are as follows:

Fire Class A: This class, represented by a green triangle, specifies use for common fires that burn solid fuels like wood. To help you remember this type, substitute "ash" for "A." Water-based extinguishers should not be used in server rooms; they are exclusively appropriate for Class A fires.

Fire Class B: This sort of definition, shown by a red square, is used for flowable liquid and gas filters. Since butane is a very flammable gas, I like to associate B with "butane" to help me remember this.

Fire Class C: This sort of brake is used for electrical fires, such as when an outlet is overloaded, and is denoted by a blue circle. To help you remember this kind, think of the letter C as "copper," as in copper electrical wiring. The ideal handheld extinguisher to use in the event of a server room fire is the multifunctional BC extinguisher (CO_2) unless you have a particular hazard system, which is not a smart idea. In a server room, electrical fires are the most prone to occur.

Fire Class D: This sort of definition is used for flammable metal fires, such as lithium, magnesium, and titanium, and is identified by a yellow decagon. In the unlikely event that a laptop's batteries catch fire, a Class D extinguisher works well. One of them should undoubtedly be available in chemical and PC repair labs. Metal fires can spread swiftly and readily to become common fires. There are very few fire extinguishers that are not red as standard; they are typically yellow. Regarding the usage of extinguishers in a crucial sector of your business, this is the only other exception. I like to associate D with "deviate" to help me remember it because of those two factors.

Fire Class K: This type, which is represented by a black hexagon, is for cooking oil filters. One kind of extinguisher that every kitchen should have is this one. If your company has a cafeteria with culinary appliances, this is crucial. To recall this style, think of the letter K for "kitchen."

Although it is not an official standard, the majority of US fire extinguisher manufacturers employ the previously listed bulleted list. There's a chance that other nations have somewhat different systems.

The ABC extinguisher is the most common fire extinguisher in buildings, but it's messy and can cause computer component corrosion. For server rooms, carbon dioxide (CO_2) extinguishers are used, which release oxygen, heat, and fuel. CO_2 is safe but can cause electrostatic discharge and suffocation. If budget allows, an ABC-rated Halotron extinguisher or a hazard protection system is recommended.

Older extinguishants like halon are no longer used due to environmental concerns. Governments have banned their use, and newer extinguishers like Halotron or FE-36 use environment-safe halocarbon agents, making them safe for humans and IT equipment.

Sprinkler Systems

A fire sprinkler system is a pressurized water supply system that delivers a large amount of water to a building through a piping distribution system. It uses sprinkler heads with glass bulbs or metal links to break when heated. The system is controlled by a valve assembly in the building's basement. Dry pipe systems are needed in colder areas where water can freeze.

An organization should periodically hold fire drills, regardless of the system, to mimic a real fire and sprinkler system activation. The security administrator should then mimic the steps outlined in Chapter 16 for disaster recovery.

Local municipalities mandate organizations to have a sprinkler system covering all floor space. However, standard wet or dry pipe systems are not suitable for server rooms due to potential damage to equipment. A pre-action sprinkler system, similar to a dry pipe system, requires specific conditions like heat or smoke to be set off.

Special Hazard Protection Systems

Protect your organization's data by installing a clean agent fire suppression system, such as Halotron and FE-36. These extinguishers leave no residue after fire extinguishing, reducing damage to computer systems and networking equipment. They are rated as ABC, preventing electrical fires and ash fires, which can cause computer failure.

The server room should be equipped with a clean agent system, such as FM-200, to protect against fires. This gaseous system uses a large tank to store fire extinguishing, which is sprayed from nozzles in the ceiling. It can quickly extinguish most fires without damaging equipment. However, these systems often use a loud alarm, requiring personnel to leave. Regular fire suppression tests and drills are crucial to ensure personnel know what to do in case of a fire. IT personnel should also simulate disaster recovery procedures if necessary.

HVAC

HVAC stands for heating, ventilation, and air conditioning, and it's crucial for data centers, server rooms, and other tech-related spaces in your facility. Servers have hot CPUs, which can cause the case's temperature to rise dramatically. It is necessary to release and exhaust this heat outside the casing. Your server room will burn up from the sheer amount of heat generated by servers and other networking equipment!

Organizations construct a powerful air conditioning system specifically designed for the server room to alleviate the problem. This can give the servers a suitable ambient temperature. The system frequently has humidity control as well. Static electricity is our adversary, as we all know. We reduce the possibility of ESD and the accumulation of static electricity by raising the humidity. Additionally, by doing this, we can prevent our equipment from being overly humid, which can also lead to failure. This system has to be on a suitably rated, dedicated circuit of its own.

To ensure the proper functioning of AC systems, it's crucial to locate the device and pipes away from servers and equipment, or use a pipeless system, with controls within the server room and monitored by authorized personnel.

To improve heat conditions, install hot and cold aisles in IT equipment. For example, a row of servers on a raised floor can be divided into a cold aisle and a hot aisle, where the heat is exhausted behind the row. The hot air is then exhausted through the floor or ceiling exhaust ducts.

In a server room, a heating system is rarely required unless the building housing the organization is located in an extremely cold climate. This is because all servers emit a lot of heat and are often in operation around the clock.

In case of a power failure, it may be necessary to temporarily shut down all but the most essential systems. Industrial control systems like SCADA can help monitor HVAC systems and their power consumption. SCADA combines hardware monitoring devices with software, allowing real-time monitoring. However, SCADA systems are vulnerable to viruses, so updating the software and securing physical interfaces, displays, and sensors is crucial.

HVAC systems require qualified professionals for maintenance and repair, often involving building management or organization. The exam should focus on HVAC systems' data availability, as IT knowledge is typically beyond their expertise.

NOTE: ANT, a proprietary wireless sensor technology, is making its way into industrial applications, health-related gadgets, home automation, and more. Originally, it was included in fitness and sports sensors. Be on the lookout for proprietary technologies that have wireless communication capabilities and design ways to identify and block them from entering the network of your company.

Shielding

As we've already shown, RFI and EMI can tamper with acceptable signals and perhaps even produce unwanted emissions. These issues can be avoided with the aid of shielding. Even though these have already been briefly covered, let's go over them in greater depth using a few examples:

Shielded twisted-pair (STP) cable: The levels of interference on the cable segment are decreased when you use STP cable because it creates a shield around the wires inside the cable. This may be helpful for computers experiencing sporadic data loss.

HVAC shielding: You can shield air conditioners and other such equipment by putting a shield around them, which keeps the EMI that the equipment generates inside the shield.

Faraday cage: Faraday cages come in various varieties. Basic examples of screened cables include TV coaxial wires. Another example would be aluminum foil-lined booster bags.

However, a room as a whole is typically referred to as a "Faraday cage." Electromagnetic energy cannot travel through walls in either direction if the entire space is protected. Therefore, if someone tries to use a cell phone within the cage, it won't work correctly because the signal can't get through the cage walls and can't be picked up by the phone from a cell phone tower. More importantly, equipment outside the cage that generates electromagnetic fields, such as wireless access points, motors, and cell phones, cannot interfere with equipment inside the cage that is sensitive to electromagnetic fields.

Shielding effectively limits interference and is common in data centers and advanced computer rooms. The Transient Electromagnetic Pulse Emanations Standard (TEMPEST) is the pinnacle of shielding technology and research. It focuses on compromising emissions and the strictness of shielding necessary for specific environments. TEMPEST-certified buildings prevent wireless devices from hacking and damage caused by high-energy electromagnetic pulses (EMPs). Although the chances of a high-energy EMP occurring near a facility are rare, some organizations and government facilities require protection.

Vehicles

Vehicles used by an organization could include cars, trucks, boats, planes, drones, tractor-trailers, and more. Since they all have built-in computers these days, it goes without saying that they can all be hacked. Additionally, the driver of the car will most likely have a mobile device with them, which may or may not communicate with the car. We have just cracked open a can of mobile worms.

But don't worry—most attacks and problems that may arise may be avoided by using the principles in this book, as most of them are similar.

Many vehicles have Apple Car Play or Android Auto, allowing seamless integration with smartphones or tablets. Disabling this technology can pose security risks, as malware can transfer to automotive computers. The Controller Area Network (CAN) bus allows communications between control units, but poorly designed systems can lead to vulnerabilities. With numerous auto manufacturers and models, the risk level increases.

Many organizations opt for fleet vehicles without in-dash computers or projection systems to reduce risk and save money. Vehicles may have SD card slots for updating GPS or automotive software, but infected cards can corrupt GPS systems. Vehicles can be disabled with passcodes or passwords.

Manufacturers and organizations use an air gap as a network security measure to isolate an entity, such as a CPU or entire network, from everything else. This method is common in industrial control systems, mission-critical controls, and military and government scenarios. The air gap can be classified as classified (secure or high side) or unclassified (insecure or low side). Data transfer between the two systems is stricter and may require the physical moving of data. Offline certificate authority (CA) is a key component in enhancing air gap security.

Vehicles should have appropriate locking systems and authentication methods like passcodes, proximity keys, and biometrics to protect internal systems. Wireless systems like Wi-Fi and Bluetooth can be vulnerable. Use prevention methods, disable unnecessary devices, use passcodes for Bluetooth, and consider geofencing to reduce Wi-Fi power levels.

Unmanned aerial vehicles (UAVs) have become increasingly common in security and defense applications. However, the risk associated with UAV technology is double-edged, as it can be exploited by organizations and attackers. Organizations can prevent this by using best coding practices, encryption, mutual authentication, and UAV-specific security standards. Additionally, well-funded attackers can use UAV technology for reconnaissance, spying, or wireless network access. Preventative measures include geo-fencing policies, strong Wi-Fi encryption protocols, physical security methods, and no-fly-zone policies.

The main takeaway is to integrate your organization's security policies and processes into every vehicle that is used. This mindset, which permeates the entire book, is that anything that has a CPU and memory is a computer and that any computer can be hacked with enough time and effort. Finally, given this definition of "computer," it is acceptable to argue that they are present almost everywhere. Handle them appropriately.

CHAPTER 18:

ACRONYMS	LIST OF ABREVIATIONS
ISO	*International Organization for Standardization*
IEC	*International Electrotechnical Commission*
HIPAA	*Health Insurance Portability and Accountability Act*
SOX	*Sarbanes-Oxley Act*
GRC	*Governance, Risk, and Compliance*
PII	*Personally Identifiable Information*
AUPs	*Acceptable Use Policies*
SLA	*Service-Level Agreement*
BPAs	*Business Partner Agreements*
MoUs	*Memorandums of Understanding*
ISAs	*Interconnection Security Agreements*
IT	*Information Technology*
UEFI	*Unified Extensible Firmware Interface*
BIOS	*Basic Input/Output System*
NIST	*National Institute of Standards and Technology*
IR	*Incident Response*
FTKs	*Forensic Tools Kits*
OOV	*Order of Volatility*
RAM	*Random Access Memory*
SCAP	*Security Content Automation Protocol*
ERP	*Enterprise Resource Planning*

Chapter 18

Policies and Procedures

Key Topic!

This chapter focuses on procedures and policies for the Security+ exam, focusing on larger concepts concerning organizations and employees. It is less tech-oriented and more people-oriented, allowing for a more comprehensive understanding of the roles of security personnel. Companies of all sizes create their policies adopt those from other organizations, or apply for company-wide standards certification from organizations like ISO. Policies are designed to protect employees and make organizations more productive and efficient. Procedures are more specific and can be interpreted more loosely, with standard operating procedures being tighter. An IT security framework is essential for organizing these procedures and policies, serving as a blueprint for an organization's security goals.

Legislative and Organizational Policies

This section discusses legislative laws and policies affecting privacy and security standards for the Security+ exam. Organizational policies are crucial for the exam, defining data classification, employee behavior, and IT equipment disposal. These policies are short, clear, and open-ended, with procedures detailing implementation. The wording may be tailored to the organization's feel and resources, and the authentication scheme used may be complex. The policy is open-ended, allowing for change over time.

Data Sensitivity and Classification of Information

Sensitive data is information that can cause a loss of security or advantage to a company if accessed by unauthorized persons. ISO/IEC 27002:2013 is a security standard that helps companies classify their data. By setting data roles such as owner, custodian, and privacy officer, and enforcing policies that govern access to information, a company can limit its exposure to security threats. Different organizations classify data in various ways, such as high, medium, and low classifications. The incorporation of these types of classifications is a key element in the multilevel security of Trusted Operating Systems (TOSs). Data handling is the process of ensuring that research data is stored, archived, or disposed of securely during and after a research project. Companies must also comply with specific laws and bills regarding data breach notification. SOX and HIPAA have sparked the concept of governance, risk, and compliance (GRC), which deals with continuous security monitoring of information system management, risk management processes, and compliance with stated requirements.

Personnel Security Policies

Organizations have policies governing employees, varying in scope from small companies to large corporations certifying to ISO 9001:2015 and ISO 9001:2008. These policies cover all aspects of the business and require rigorous examination and accreditation by an accrediting body. IT professionals focus on security policies, working with procedural, technical, and outside consultants. While security should be a priority, actual work should also be prioritized. Understanding and implementing these policies is crucial for maintaining a secure work environment.

Privacy Policies

The Privacy Act of 1974 sets standards for PII security, but organizations often define their privacy policies. These policies outline how users' identities and other information will be secured, including access, retention, and distribution. They also govern data labeling to ensure employees understand the data they are handling and prevent mishandling of confidential information. Systems administrators should consult the privacy policy before gathering user information.

Quick Hits!

Acceptable Use

Acceptable use policies (AUPs) outline rules for using technology infrastructure in an organization. They state what users are allowed to do and are often signed by employees before beginning work on systems. AUPs protect the organization and outline what employees should and should not be working on. Unauthorized data sharing is prohibited, and employees must not take information or equipment home without permission. Strong data ownership policies are needed, identifying which portion is owned by the organization and the employee, and backing up any organizational data on mobile devices.

Change Management

Change management is a structured approach to altering computer systems, networks, or IT procedures. It involves introducing changes to department heads, approving them before implementation, and monitoring and documenting the process. Network engineers must understand the importance of each part of the IT infrastructure. For instance, modifying an unused interface on a firewall without consulting the right team can be negligent. Forms requiring the change's reason and signature by multiple people can slow progress but ultimately benefit the network engineer. In larger organizations requiring certifications like ISO 9001:2015, change management can be complex. IT personnel should have charts of personnel and department heads, and current procedures should outline who needs to be contacted for proposed changes.

Separation of Duties/Job Rotation

Separation of duties involves assigning multiple people to a task and distributing control over a system. Job rotation enforces this by allowing regular role changes, increasing user insight and skill level, and reducing fraud risk. These policies enhance organizational security by limiting control and increasing employee knowledge.

Mandatory Vacations

Organizations may mandate employees to take X consecutive days of vacation annually, such as an IT director taking five consecutive days to force another person into their role. This is to prevent malicious activities like fraud, sabotage, and embezzlement, as IT professionals often access networks remotely.

Quick Hits!

Onboarding and Offboarding

Onboarding is the process of adding a new employee to an organization's identity and access management system, often through training, meetings, and handbooks. It is a socialization technique that improves job performance and satisfaction. Offboarding, on the other hand, involves removing an employee from a federated identity management system, restricting rights, and conducting exit interviews. These processes can also apply to business partners, who may have access to shared data through technologies like extranets and community clouds.

Due Diligence

Due diligence in information security involves identifying and managing IT infrastructure risks, assessing vulnerabilities, and implementing policies to address them. It also involves studying company history, failures, errors, and user issues.

Due Care

Due care is a mitigation strategy implemented by an organization to protect against risks discovered during due diligence.

Due Process

Due process is the principle that organizations must uphold and protect the rights of their employees to prevent state interference and frivolous lawsuits.

Quick Hits!

User Education and Awareness Training

Organizational policies require employees to understand and adhere to them. HR personnel should be trained in guidelines and enforcement, while security awareness training is an ongoing process with varying types and roles. Employees should be trained on personally identifiable information (PII) and password policies, as well as account changes. IT personnel should be familiar with the organization's password policy and lockout policy. Additional privacy training, HIPAA training, or other security awareness training may be necessary depending on the organization. Role-based training can be used to assess the effectiveness of training and validate compliance with policies. Instructors may use learning management systems and training metrics to gauge the effectiveness of training and analyze trainees' security posture. Overall, employee training is crucial for maintaining a secure and compliant organization.

Quick Hits!

How to Deal with Vendors

Before signing any contracts, organizations should consult with an attorney to ensure a clear understanding of their vendor policies and the level of agreement between them. A proper service-level agreement (SLA) can help define the vendor's and organization's responsibilities, performance expectations, and timeframes for repair. An attorney should be involved in the review process to ensure the SLA covers all required points.

Business partners agreements (BPAs) establish the profits, responsibilities, and exit strategies for partners. Blanket purchase agreements (BPAs) are common in government contracts but should have an ending for the contract length. Some organizations may use memorandums of understanding (MoUs) to establish convergence in ordering services.

Interconnection security agreements (ISAs) are agreements between organizations that own and operate connected IT systems and data sets. ISAs document the technical and security requirements of the interconnection, which is more precise than an SLA.

An SLA is a contract between a service provider and a customer that specifies the nature of the service and the level of service offered. A BPA does not have inherent security planning, and an MoU is an understanding between two organizations or government agencies. Understanding these terms is essential for interacting with business people and understanding special contractual relationships between organizations.

How to Dispose of Computers and Other IT Equipment Securely:

Organizations can recycle or donate IT equipment, but it is important to consult the organization's policy regarding computer disposal and local municipal guidelines. A basic recycling policy and procedure might look like this:

A must Know!

1. Define the equipment to be disposed of.
2. Obtain temporary storage for the equipment.
3. Have the appropriate personnel analyze the equipment.
4. Verify whether the equipment is outdated and can be used elsewhere in the organization. If a device can be used in another area, format it, flush it, or reset it to the original default.
5. Sanitize the devices or computers.
6. Check for removable media inside or connected to the computer.
7. Remove RAM, label it, and store it.
8. Remove the hard drive, sanitize it, and store it. If necessary, pulverize or destroy the device.
9. Reset any UEFI/BIOS or other passwords to the default setting.
10. Recycle or donate items as necessary.

Data removal is the most important element of computer recycling. Companies typically employ one of three options when faced with the prospect of data removal: clearing, purging, or destruction. Clearing involves the removal of data with a certain amount of assurance that it cannot be reconstructed while purging involves physically destroying the storage media through various methods.

The type of data removal used will depend on the data stored on the drive. If there is no personally identifiable information, it may be cleared and released outside the company.

Incident Response Procedures

Incident response is a set of procedures that an investigator follows when examining a computer security incident. It is part of computer security incident management, which involves monitoring and detecting security events on a computer network and executing proper responses to those security events. IT employees often discover or act as investigators, depending on the organization's resources and budget. IT personnel needs to be well-briefed on policies regarding the reporting and disclosure of incidents.

Incidents should not be confused with events, as they are separate entities. An example of an incident would be multiple DDoS attacks launched at an organization's web servers over a workday. Organizations may use varying incident response processes, with the National Institute of Standards and Technology (NIST) breaking the process down into four main phases. The CompTIA Security+ objectives include six phases, with some organizations breaking it down into ten.

The CompTIA Security+ objectives' incident response plan (IR) can be summarized as follows:

Phase 1: Preparation: This involves preparing for an event, such as a data breach, and ensuring that the organization has a well-planned incident response procedure, a strong security posture, and a knowledgeable chief information security officer (CISO).

Phase 2: Identification: Recognizing whether an event should be classified as an incident and determining whether to make contact with other groups or escalate the problem if necessary.

Phase 3: Containment: Isolating the problem, such as removing the attacker from a padded cell or physically disconnecting devices from the network.

Phase 4: Eradication: Removal of the attack or threat, quarantine of computers, device removal, and other mitigation techniques.

Phase 5: Recovery: Retrieving data, repairing systems, re-enabling servers and networks, reconstituting server rooms and/or the IT environment. Damage and loss control come into play in this phase, and affected systems should be monitored for any repercussions.

Incident response policies can be more in-depth and specific, but common practices and standards for incident response can be found in ISO/IEC 27002:2013 or NIST Special Publication 800-61 Revision 2.

An incident response policy typically outlines the initial incident management process, including identification of the problem, tracking tickets, and change controls. It defines first responders who analyze incident data and determine the severity of the incident. If an incident is considered an emergency, it outlines the escalated process, coordinator, lockdown procedures, containment, system repair, and further investigation procedures.

The policy also covers the collection and preservation of evidence, such as log files, alerts, and captured packets, and the preservation of forensic images. Modifications during the investigative process can invalidate evidence validity in court. Establishing a chain of custody, or paper trail of evidence, is crucial for preserving evidence. Legal holds can affect the procurement of evidence, and the policy lists proper procedures for procurement.

The policy also covers damage and loss control, including stopping the spread of damage to other IT systems and minimizing data loss. The toughest part of the job involves hardcore forensics, which may be handled internally or by third-party vendors like forensics consultants and specialists. The policy may dictate how computer forensics should be carried out, including deciphering information from hard disks and identifying tampering by outside entities. It also specifies rules for investigating an attacker's actions, such as verifying data integrity. Proper computer forensics investigations are essential in case legal action is taken.

Forensic procedures in incident response involve capturing and hashing system images, analyzing data with software tools, capturing screenshots, reviewing network traffic captures and logs, and capturing video surveillance equipment. These procedures are typically applied during the containment phase and can be performed during other phases as well.

Capture and hash system images: If a computer's data is to be used as evidence, the entire drive should be imaged before investigation. The imaging process should be secure and logged, and the image itself should be hashed. The order of volatility (OOV) is important when imaging any media, and LiveCDs, LiveDVDs, and flash drives are commonly used for this purpose.

Forensic tools (FTKs) can be invaluable for analyzing data files. Screenshots should be taken directly from the affected computer rather than using screen-capturing software installed on the affected computer. Network traffic captures and logs should be preserved, hashed, and stored, including firewall logs, server logs, and router/switch logs.

Capture video: Any video surveillance equipment that recorded an incident should be analyzed. The process should be secured and logged to prevent tampering. The "real" time should be compared to the video's time stamp, known as the record time offset.

Order of volatility (OOV) is crucial when analyzing captured data, as information stored in memory, cache, or CPU registers only lasts for seconds. Witness statements are also essential for court cases and investigations.

To prepare for a computer seizure, follow these steps: 1) Photograph the computer and scene, 2) Do not turn the computer on if off, 3) Photograph the screen if on, 4) Collect live data from the RAM image, 5) Collect a logical image of the drive if encrypted, 6) Unplug the power cord, 7) Diagram and label all cords, 8) Document device model numbers and serial numbers, 9) Disconnect all cords and devices, 10) Collect an image of the hard drive using a hardware imager or software tools, 10) Hash the image, 10) Package components using antistatic evidence bags, 10) Collect additional storage media, 10) Keep media away from magnets and radio transmitters, 10) Collect instruction manuals, documentation, and notes, and 6) Document all steps performed during the seizure. Training personnel is crucial for quick and efficient incident analysis and preventing disasters. Need-to-know security concepts help employees quickly execute necessary actions and secure confidential or top-secret information in classified environments.

IT Security Frameworks

An IT security framework is a crucial tool for organizations to provide implementable security controls for their IT environment and document the processes, procedures, and policies used to implement them. Organizations can either create their framework or use widely accepted standards, such as ISO/IEC 27000, NIST, ISACA's COBIT, the Information Technology Infrastructure Library (ITIL), Business Information Services Library (BiSL), and Project Management Body of Knowledge (PMBOK). These standards can be regulatory or nonregulatory and are typically used in the United States or other countries. Reference frameworks can also be industry-specific, defining how precise tasks and problems within an organization should be approached. For example, a company might require a secure configuration guide for mobile devices or benchmarking servers. A well-thought-out IT security framework can also include risk analysis and vulnerability assessment tools, such as the Security Content Automation Protocol (SCAP), and how to properly utilize enterprise resource planning (ERP) software. The goal is to organize a group of processes, procedures, and policies into a single cohesive agenda that all employees can easily understand and work within.

Installing the SSD card

1. Insert the SSD card into the connector on the system.
2. Replace the M2*3 screw that secures the SSD card to the system.
3. Place the Mylar shield over the SSD.

Installing hard drive

1. Insert the hard drive into the slot on the system.
2. Replace the four (M2 x 2.7) screws to secure the hard drive to the system.

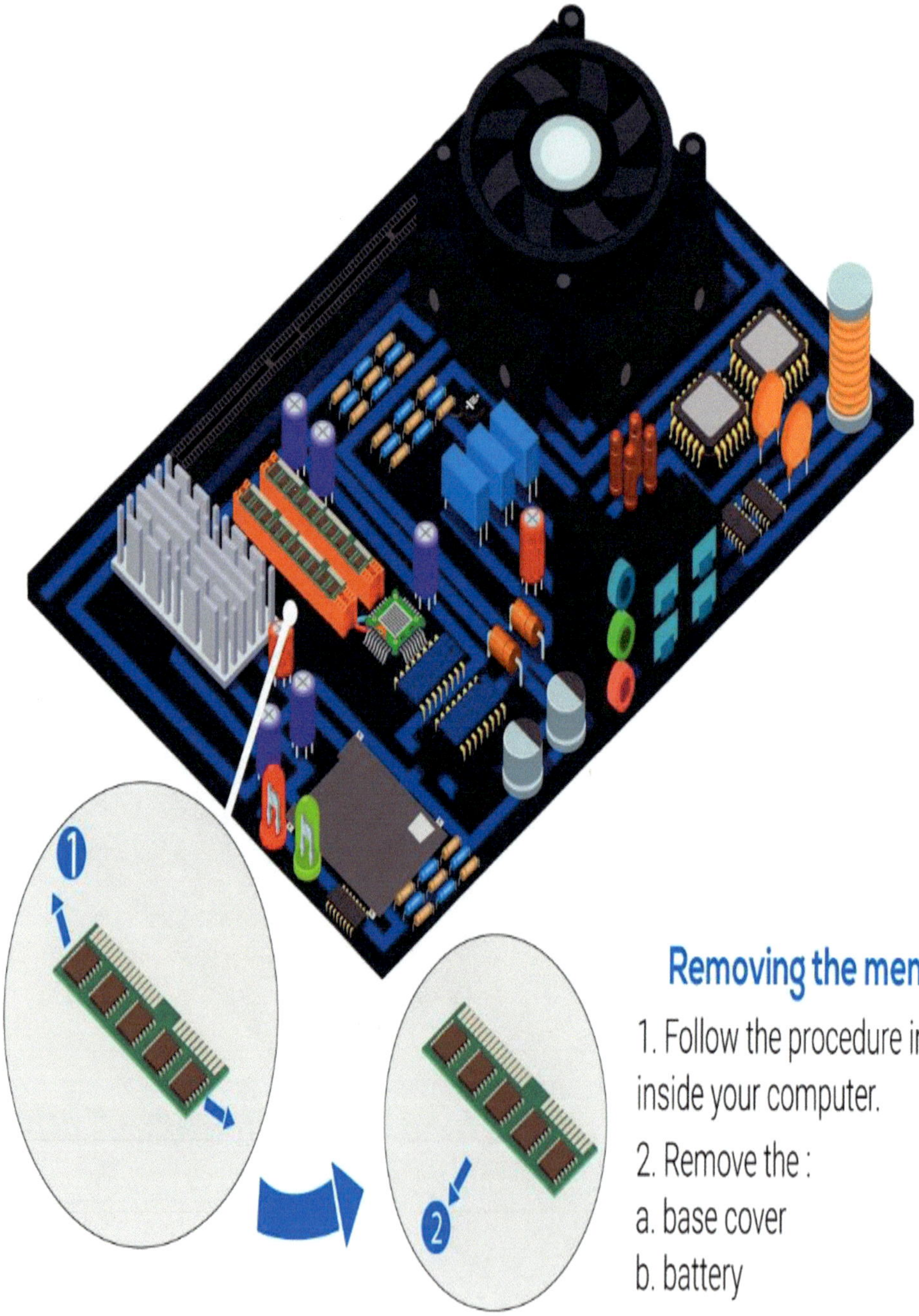

Removing the memory module

1. Follow the procedure in Before working inside your computer.
2. Remove the :
a. base cover
b. battery

Installing WLAN card

1. Insert the WLAN card into the connector on the system board.
2. Connect the WLAN antenna cables to the connectors on the WLAN card
3. Place the WLAN card bracket to secure the WLAN cables.
4. Replace the M2x3 screw to secure the WLAN card to the system.

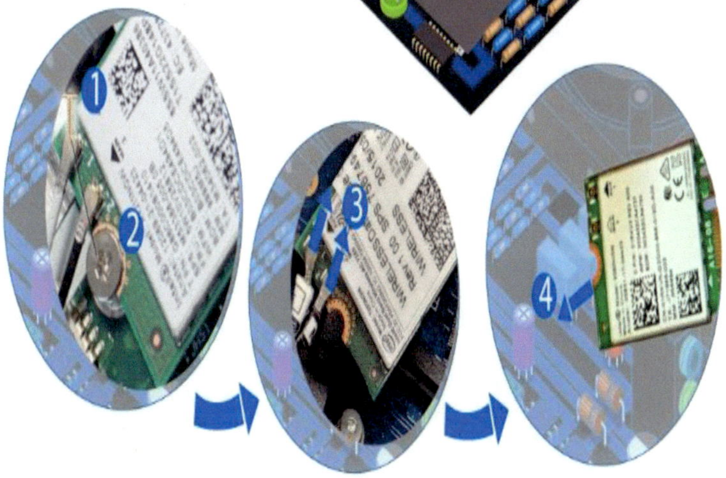

Installing coin cell battery

1. Affix the coin cell battery on the system board.
2. Connect the coin cell battery cable to the connector on the system board

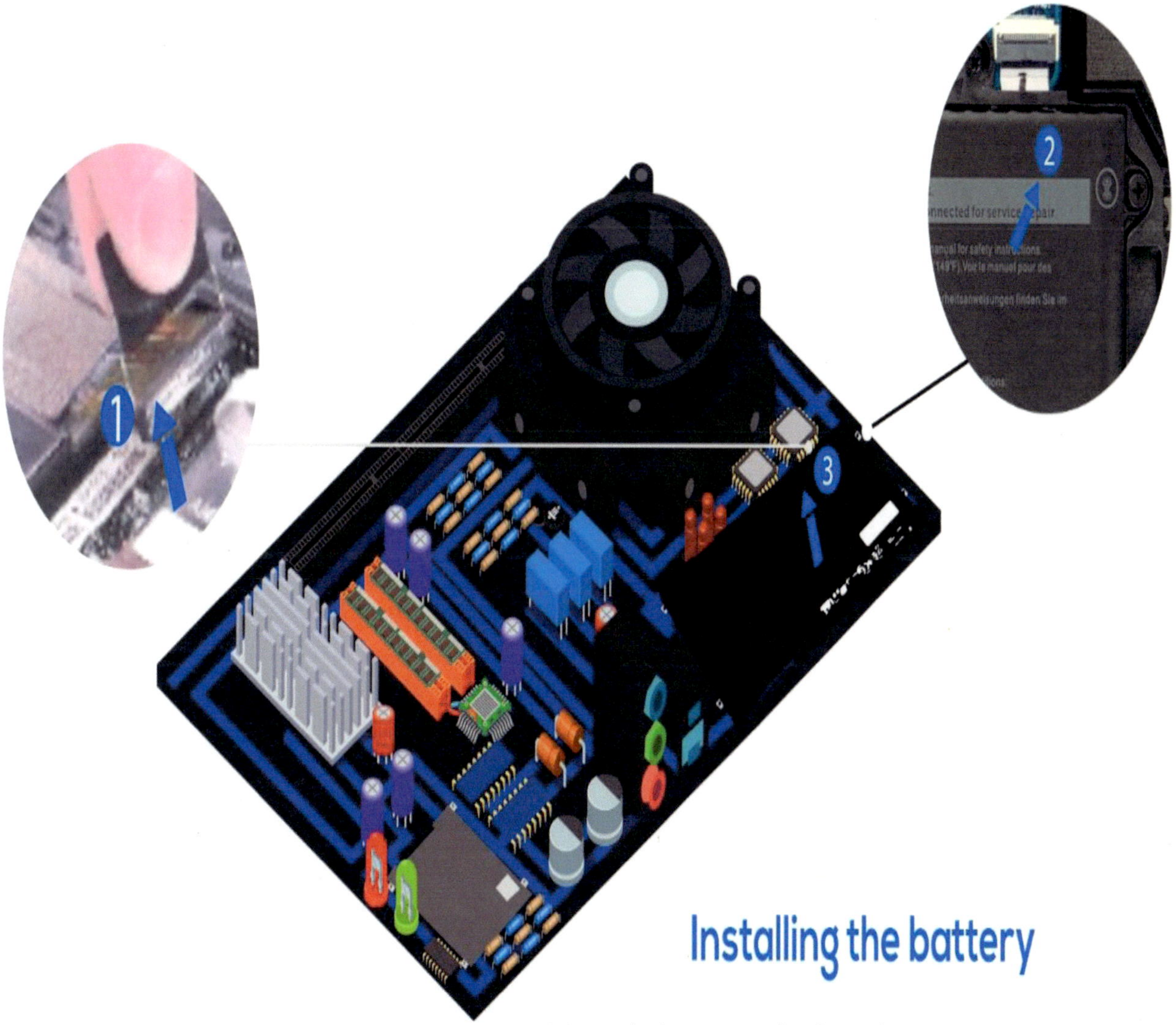

Installing the battery

1. Insert the battery into the slot on the system.
2. Route the battery cable through the routing channel.
3. Tighten the M2x6 captive screw to secure the battery to the system.

Made in the USA
Las Vegas, NV
03 November 2026

57423312R00197